Application of the SPARROW Model to Assess Surface-Water Nutrient Conditions and Sources in the United States Pacific Northwest

By Daniel R. Wise and Henry M. Johnson

Scientific Investigations Report 2013–5103

U.S. Department of the Interior
U.S. Geological Survey

U.S. Department of the Interior
SALLY JEWELL, Secretary

U.S. Geological Survey
Suzette M. Kimball, Acting Director

U.S. Geological Survey, Reston, Virginia: 2013

For more information on the USGS—the Federal source for science about the Earth, its natural and living resources, natural hazards, and the environment, visit http://www.usgs.gov or call 1–888–ASK–USGS.

For an overview of USGS information products, including maps, imagery, and publications, visit http://www.usgs.gov/pubprod

To order this and other USGS information products, visit http://store.usgs.gov

Suggested citation:
Wise, D.R., and Johnson, H.M., 2013, Application of the SPARROW model to assess surface-water nutrient conditions and sources in the United States Pacific Northwest: U.S. Geological Survey Scientific Investigations Report 2013–5103, 32 p., http://pubs.usgs.gov/sir/2013/5103/.

Contents

Figures

Tables

Conversion Factors, Datum, and Abbreviations and Acronyms

Conversion Factors

Multiply	By	To obtain
Length		
centimeter (cm)	0.3937	inch (in.)
millimeter (mm)	0.03937	inch (in.)
kilometer (km)	0.6214	mile (mi)
Area		
square meter (m^2)	0.0002471	acre
square kilometer (km^2)	247.1	acre
square kilometer (km^2)	0.3861	square mile (mi^2)
Mass		
kilogram per year (kg/yr)	2.205	pound per year (lb/yr)
Application rate		
kilograms per hectare per year [(kg/ha)/yr]	0.8921	pounds per acre per year [(lb/acre)/yr]
kilograms per square kilometer per year [(kg/(km^2)/yr]	0.008921	pounds per acre per year [(lb/acre)/yr]
Flow		
cubic meter per second (m^3/s)	35.32	cubic foot per second (ft^3/s)

Datum

Horizontal coordinate information is referenced to the North American Datum of 1983 (NAD 83).

Altitude, as used in this report, refers to distance above the vertical datum.

Conversion Factors, Datum, and Abbreviations and Acronyms—Continued

Abbreviations and Acronyms

CLRW	Clearwater River basin
CMAQ	Community Multiscale Air Quality
HUC6	six-digit Hydrologic Unit Code
MCOL	Middle Columbia River basin
NAWQA	National water-Quality Assessment Program
NHD	National Hydrography Data set
NLCD	National Land Cover Database
NLLS	nonlinear least squares regression
NOCR	northern Oregon coast
NPDES	National Pollutant Discharge Elimination System
PNW	Pacific Northwest
RMSE	root mean squared error
SPARROW	Spatially Referenced Regression on Watershed attributes model
SPOK	Spokane River basin
TMDL	Total Maximum Daily Load
TN	total nitrogen
TP	total phosphorus
USEPA	U.S. Environmental Protection Agency
USGS	U.S. Geological Survey
WACR	Washington coast

Application of the SPARROW Model to Assess Surface-Water Nutrient Conditions and Sources in the United States Pacific Northwest

By Daniel R. Wise and Henry M. Johnson

Abstract

The watershed model SPARROW (Spatially Referenced Regressions on Watershed attributes) was used to estimate mean annual surface-water nutrient conditions (total nitrogen and total phosphorus) and to identify important nutrient sources in catchments of the Pacific Northwest region of the United States for 2002. Model-estimated nutrient yields were generally higher in catchments on the wetter, western side of the Cascade Range than in catchments on the drier, eastern side. The largest source of locally generated total nitrogen stream load in most catchments was runoff from forestland, whereas the largest source of locally generated total phosphorus stream load in most catchments was either geologic material or livestock manure (primarily from grazing livestock). However, the highest total nitrogen and total phosphorus yields were predicted in the relatively small number of catchments where urban sources were the largest contributor to local stream load. Two examples are presented that show how SPARROW results can be applied to large rivers—the relative contribution of different nutrient sources to the total nitrogen load in the Willamette River and the total phosphorus load in the Snake River. The results from this study provided an understanding of the regional patterns in surface-water nutrient conditions and should be useful to researchers and water-quality managers performing local nutrient assessments.

Introduction

The SPARROW (Spatially Related Regressions on Watershed attributes) model was used to estimate mean annual total nitrogen (TN) and total phosphorus (TP) loads in surface-waters across the Pacific Northwest (PNW) region of the United States. This modeling effort used the National Hydrography Dataset Plus (NHD) (Horizon Systems, 2013) as a hydrologic framework and serves as a refinement of SPARROW models of the PNW that used the RF1 (River Reach File 1) hydrologic framework (Brakebill and others,2011; Wise and Johnson, 2011). In addition to the hydrologic network, many of the catchment attributes that were compiled for the RF1 SPARROW models were refined for the new models. Using the NHD as the hydrologic framework for SPARROW provided predictions for a greater number of stream reaches compared to the RF1 hydrologic network. The PNW NHD SPARROW models included predictions for 232,811 stream reaches compared to 12,039 stream reaches for the PNW RF1 models.

Surface-water nutrient enrichment has been identified as a water-quality problem across the PNW (Wise and Johnson, 2011) and modeling plays a central role in water-quality management by providing a means for predicting water-quality conditions and assessing the effectiveness of proposed improvement strategies (National Research Council, 2001). The results from the PNW NHD SPARROW models will allow environmental managers and other stakeholders to identify the sources that contribute the largest amounts of nutrients to downstream waters and to evaluate nutrient reduction scenarios. The inclusion of smaller streams and headwater reaches in the model simulation results compared to the RF1 models also will allow for refined analyses of water-quality issues.

Purpose and Scope

This report describes the methods that were used to assess surface-water nutrient conditions in the PNW and to present the results from that assessment. The objectives of the assessment were (1) to calibrate TN and TP SPARROW models for the PNW using the NHD stream network; (2) to use the calibrated models to estimate mean annual nutrient conditions; and (3) to quantify the relative contribution of different nutrient sources to instream nutrient loads.

Description of the Modeling Domain

The domain of the PNW NHD SPARROW models covered about 708,000 km², but did not include the area of the Columbia River basin (about 106,000 km²) that drains the Canadian part of the river basin. In the United States, the domain covered parts of eight states and included five major regions (fig. 1; table 1): Pacific Coast, West Side Basins, Columbia River Basin, Snake River Basin, and Oregon Closed Basins, and 22 level three, six-digit Hydrologic Unit Code (HUC6) watersheds (Seaber and others, 1987). The Pacific Coast region is characterized by steep, forested watersheds that drain directly to the Pacific Ocean. The West Side Basins region lies between the Cascade Range and the Coast Range (including the Olympic Mountains in Washington). Although most of this region is forested, it contains most of the population in the PNW model domain (including Seattle and Portland) as well as large areas of agricultural land. The Columbia River Basin and Snake River Basin regions are dominated by sparsely vegetated, rocky areas, high desert steppe, semi-arid forests, and areas of intensive agricultural

production. The Oregon Closed Basins region is characterized by alternating regions of narrow, uplifted mountains, flat arid plains, playas, and alkali lakes. This is the least populated region in the PNW model domain and contains little forestry or agricultural activity. In 2001, scrub and grassland covered 43 percent of the modeling domain, forest covered 40 percent, agriculture covered 9 percent, and developed land covered 7 percent. The remaining 1 percent was various minor land cover types (Homer and others, 2004).

Table 1. Definitions for abbreviations used for six-digit Hydrologic Unit Code (HUC6) watersheds in the United States Pacific Northwest.

Region	HUC6 abbreviation	HUC6 watershed
Pacific Coast	WACR	Washington Coastal Rivers
	NOCR	Northern Oregon Coastal Rivers
	SOCR	Southern Oregon Coastal Rivers
West Side Basins	PUGT	Puget Sound
	LCOL	Lower Columbia River
	WILL	Willamette River
Columbia River Basin	KOOT	Kootenai River
	PDOR	Pend Oreille River
	SPOK	Spokane River
	UCOL	Upper Columbia River
	YAKI	Yakima River
	JDAY	John Day River
	DESC	Deschutes River
	MCOL	Middle Columbia River
Snake River Basin	SNKH	Snake River Headwaters
	USNK	Upper Snake River
	MSBS	Middle Snake–Boise Rivers
	MSPW	Middle Snake–Powder Rivers
	SALM	Salmon River
	CLRW	Clearwater River
	LSNK	Lower Snake River
Oregon Closed Basins	ORCB	Oregon Closed Basins

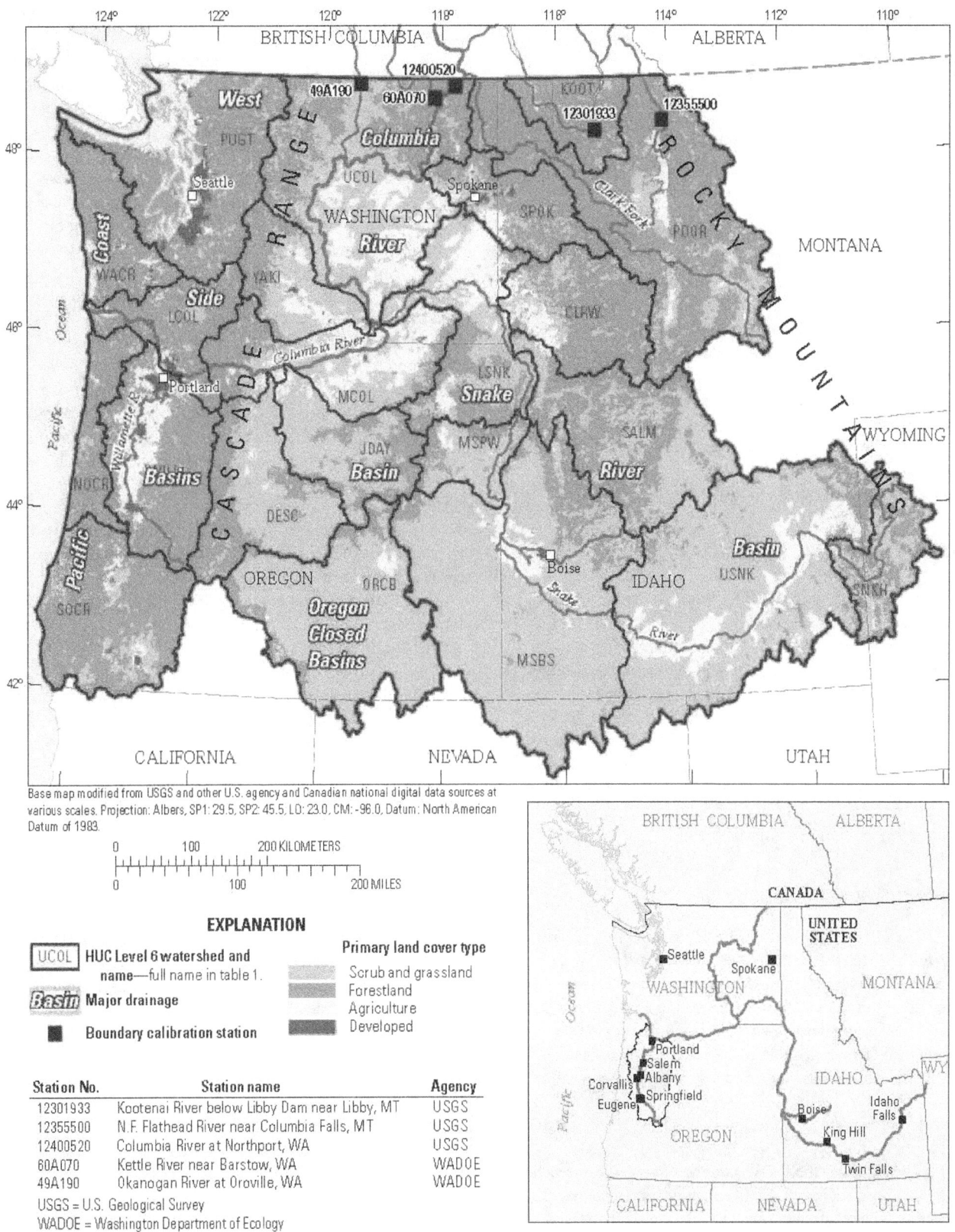

EXPLANATION

UCOL	HUC Level 6 watershed and name—full name in table 1.
Basin	Major drainage
■	Boundary calibration station

Primary land cover type

Scrub and grassland
Forestland
Agriculture
Developed

Station No.	Station name	Agency
12301933	Kootenai River below Libby Dam near Libby, MT	USGS
12355500	N.F. Flathead River near Columbia Falls, MT	USGS
12400520	Columbia River at Northport, WA	USGS
60A070	Kettle River near Barstow, WA	WADOE
49A190	Okanogan River at Oroville, WA	WADOE

USGS = U.S. Geological Survey
WADOE = Washington Department of Ecology

Figure 1. Major drainages, Hydrologic Unit Code (HUC6) watersheds, and primary land cover types in the United States Pacific Northwest.

Methods

The SPARROW Model

The SPARROW model is a hybrid statistical and mechanistic model for estimating the movement of mass through the landscape under long-term, steady state conditions (Schwarz and others, 2006). The model uses data describing catchment attributes (nutrient sources, landscape characteristics, and stream and water body properties) to explain the spatial variation in measured, mean annual stream load (expressed as kilograms per year). The measured, mean annual stream loads are the dependent variable (the calibration data set) for the model and the catchment attributes are the explanatory variables. In this report, the variables representing nutrient sources are called nutrient source terms and the variables representing the delivery of nutrients from land to water are called delivery terms. A calibrated SPARROW model can be used to predict water-quality conditions throughout a surface-water network, including areas where no water-quality data exists.

Model Input Data Sets

The input data sets used for this study were similar to those used to develop the RF1 SPARROW models for the PNW (Wise and Johnson, 2011). An important structural change from the earlier models was the use of the National Hydrography Dataset Plus (NHDPlus) as the surface-water drainage network for the models rather than the RF1 hydrologic network. Refinements also were made to many of the input data sets, including the calibration data set of mean annual nutrient loads, the estimates of nutrient sources, and the accounting of irrigation diversions, power returns, and discharge from large spring complexes.

Surface-Water Drainage Network

The NHD Plus Version 2 for Hydroregion 17 (Horizon Systems, 2013) was used to represent the surface-water drainage network in the models developed for this study. Hereafter, this data set will be referred to as the NHD. The NHD is a comprehensive set of digital spatial data that contains information about surface water features such as lakes, ponds, streams, and rivers (Simley and Carswell, 2009). The surface water features represented in the NHD largely correspond to the features on 1:100,000 scale USGS topographic maps.

The NHD for the PNW is divided into 232,811 reaches, which vary in size from small, ephemeral streams that can go years without streamflow to the Columbia River with a mean annual streamflow of 9,575 cubic meters per second near its confluence with the Pacific Ocean (U.S. Geological Survey, 2012). The NHD identifies the incremental catchment for each reach. An incremental catchment is defined as the area that drains directly to a reach without passing through another reach. Most reaches in the NHD represent streams or inland water bodies, such as lakes and reservoirs. However, some reaches represent coastlines or closed basins, which do not have a surface water connection to other reaches in the NHD. In building the hydrologic framework for the SPARROW models, reaches representing streams, inland water bodies, and coastlines were retained, but reaches representing closed basins were eliminated from the network. The NHD contains minimal information on stream reaches and catchments in Canada, but does provide sufficient information to properly route surface water into the United States.

Irrigation and power networks in the PNW divert large amounts of water from streams and reservoirs, and these diversions needed to be accounted for to properly estimate nutrient transport through surface waters. The PNW SPARROW models included a reach attribute that simulated the diversion of streamflow in the drainage network. This was done by estimating the fraction of streamflow and, therefore, nutrient load that was delivered from one reach to the reach immediately downstream (based on long term average conditions). In the SPARROW model, nutrient load that is removed because of irrigation diversions is not explicitly accounted for as return flow through the modeling network, although many of the agricultural returns in the PNW are represented in the NHD drainage network.

Calibration Data Set

The calibration data set for the models consisted of mean annual TN and TP stream loads that were estimated from water-quality data obtained from Federal agencies, State regulatory agencies, one county government, and one water pollution control district and streamflow data collected primarily by the U.S. Geological Survey (USGS). Water-quality monitoring stations were selected as TN, TP, or both calibration stations if they were close enough to a nearby streamflow gaging station and met the minimum criteria for the number of TN and TP samples (20), seasonal coverage (3 samples per season), and period of record (last sample collected no later than 1995 if there were at least 5 years of data or last sample collected no later than 1999 if there were less than 5 years of data). The mean annual TN and TP stream loads were estimated using the USGS Fluxmaster model (Schwarz and others, 2006), which relates the loads measured at water-quality monitoring stations (the calibration stations) to measured streamflow, season, and time. There were 179 calibration stations where TN loads were estimated, 220 where TP loads were estimated, and 177 where both TN and TP loads were estimated. This resulted in a total of 222 calibration stations. A breakdown of the calibration stations by agency and location is provided in table 2.

Table 2. Sources of water-quality data used to estimate total nitrogen and total phosphorus calibration stream loads for the National Hydrography Dataset SPARROW models developed for the United States Pacific Northwest.

Level	Agency	Number of stations
Federal	U.S. Geological Survey	84
	Bureau of Reclamation	12
State	Oregon Department of Environmental Quality	58
	Washington Department of Ecology	55
Local	King County, Washington	9
	Clean Water Services, Oregon	4
Total		222

Location	Number of stations
Washington	84
Oregon	74
Idaho	51
Montana	9
Wyoming	4
Total	222

The mean annual TN and TP stream loads were detrended to 2002 to account for differences in record length, hydrologic conditions, and sample size among the calibration stations (Preston and others, 2009). Differences between the calibration stream loads used in the PNW NHD SPARROW models and those used in the PNW RF1 SPARROW models (see Saad and others, 2011) were a result of closer examination of the water-quality and streamflow data used to estimate the loads. In one case, however, the difference was due to the inclusion of a water-quality monitoring station that was located on an NHD stream reach not represented in the RF1 network.

Catchment Attribute Data

The catchment attribute data for computing explanatory information used in the PNW NHD SPARROW models consisted of nutrient sources, boundary loads, and land-to-water delivery factors.

Nutrient Sources

All nutrient sources considered in the PNW RF1 SPARROW models were considered in the PNW NHD SPARROW models. Fixation of nitrogen from the atmosphere in forested areas was represented by the area of forestland, and the weathering of geologic phosphorus was represented by the area of forestland and the area of scrubland plus grassland. The approach used to represent geologic phosphorus was similar to that used in other SPARROW applications (Smith and others, 1997; Alexander and others, 2008). Other nutrient sources considered for use in the models were (1) the discharge from permitted wastewater treatment facilities including fish farms and hatcheries (point sources), (2) the area of developed land to represent nonpoint urban sources, (3) the number of people living in areas not served by municipal sewage districts to represent nitrogen leaching from septic tanks, (4) the atmospheric deposition of nitrogen, (5) the application of farm and nonfarm fertilizer, (6) the application or deposition of livestock manure, and (7) the leaching of nitrogen from red alder trees (*Alnus rubra*). The methods used to estimate these nutrient sources are described in appendix A.

The TN and TP NHD models also accounted for nutrient loads from the largest spring complexes, which are collections of natural springs that discharge into or near a stream and contribute a substantial amount of the flow in that stream, and the nutrient loads associated with the return of water from off-stream power generation facilities. Spring complexes and power returns are not sources of nutrients, but are pathways for the return of nutrients to a stream. They are represented in the models as point sources because there is no mechanism within the NHD network or in the SPARROW model to accommodate these unique pathways for nutrient movement from the landscape to the stream. The methods used to estimate these loads also are included in appendix A.

Boundary Loads

Five of the calibration stations were on stream reaches with large upstream drainage areas that were primarily in Canada (Kettle River, Okanogan River, Columbia River, North Fork of the Flathead River, and Kootenai River; fig. 1). Because catchment attribute data for computing explanatory information in the Canadian part of the modeled area were not available, these five calibration stations were used as boundary conditions for the models. The TN and TP models were configured so that the load entering the stream network at these boundary reaches was equal to the measured mean annual TN and TP load and that the processes occurring upstream of and within their incremental catchments had no effect on the calibration of the TN and TP models. There also were 184 incremental catchments that included some Canadian land but that did not drain to a boundary reach. Incomplete attribute data for these catchments was expected to have little influence on the model calibrations, however, because they represented 0.23 percent of the area of the modeling domain.

Land-to-Water Delivery

The delivery of nutrients from land to water was modeled by considering land cover, climate, soil properties, geology, and hydrology. Most of these landscape properties were compiled by the National Water-Quality Assessment Program (NAWQA) as part of a national effort and were summarized for each incremental NHD catchment (Michael Wieczorek, U.S. Geological Survey, written commun., June 11, 2011), and two landscape properties were compiled specifically for the PNW NHD SPARROW models (mean annual solar radiation and the extent of arid land irrigation). The methods used to compile the two data sets representing mean annual solar radiation and the extent of arid land irrigation are described in appendix A.

Nutrient Loss in Free-Flowing Streams and Impoundments

The SPARROW model can simulate the net effect of processes that lead to permanent nutrient loss (particulate settling and benthic denitrification) within surface waters. When modeling mean annual conditions an assumption is made that there is no net gain or loss of nutrients due to the growth and decay of aquatic plants (Schwarz and others, 2006). Nutrient loss in free-flowing streams is modeled in SPARROW using a first-order decay formulation that is a function of the time of travel for each reach (reach length divided by estimated mean annual velocity). As a result, the stream loss coefficient in the model is expressed as day^{-1}. Estimates of different instream nutrient-decay rates for different stream categories can be specified in the SPARROW model. In most SPARROW models developed for other parts of the United States, these stream categories were based on a gradient of mean annual streamflow, but they also can be based on stream type (for example, perennial or intermittent), temperature, or other measures that affect nutrient uptake and removal. Nutrient loss in impoundments such as lakes and reservoirs is modeled in SPARROW as an apparent settling velocity that is expressed in units of length per time and is a function of the areal hydraulic load (estimated mean annual streamflow through an impoundment divided by the surface area of the impoundment). As a result, the impoundment loss coefficient in the model is expressed as meters per year. All of the information needed to compute the parameters used to estimate nutrient loss was included with NHDPlus.

Model Calibration

The explanatory variables included in the TN and TP models represented statistically significant or otherwise important geospatial variables. The significance of the coefficients for each of the nutrient source terms (which were constrained to be positive) were determined by using a one-sided t-test and a significance level of 0.10. The significance of the coefficients for each of the delivery terms (which were allowed to be positive or negative, reflecting either enhanced or attenuated delivery, respectively) was determined by using a two-sided t-test and a significance level of 0.05. The significance of the coefficients for the variables representing nutrient loss in free-flowing streams and impoundments (which were constrained to be positive) was determined by using a one-sided t-test and a significance level of 0.10. Final model selection was based on the overall model fit by evaluating the yield R-squared (R^2) and the root mean squared error (RMSE), and by evaluating the residuals for spatial patterns. The yield R-squared is the R-squared value for the natural logarithm of yield and is considered a better measure of goodness of fit than R-squared because it accounts for the effect of contributing area, which can explain much of the variation in stream load. The spatial patterns in model fit were evaluated by calculating and mapping the studentized residual for each calibration station. The studentized residual is equal to the model residual (the difference between the natural logarithm of measured load and predicted load) divided by an estimate of its standard deviation.

The SPARROW model uses a weighted nonlinear least squares (NLLS) regression to estimate model coefficients and provides a way to assess uncertainty in these estimated coefficients. Because of the nonlinear manner in which the estimated coefficients enter the model, this uncertainty needs to be evaluated using a bootstrap resampling method (Schwarz and others, 2006). The method is implemented through repeated estimation of the SPARROW model (200 times for these applications) to obtain a range of values for each coefficient, from which a mean value (the nonparametric bootstrap estimate) is estimated. The overall stability of each of the models was evaluated by comparing the NLLS estimates of the model coefficients to the nonparametric bootstrap estimates. The 90 percent confidence intervals for the NLLS coefficients in each model were generated by using the standard errors and a t-distribution with N-k degrees of freedom, where N was the number of calibration sites and k was the number of coefficients.

Analysis of Model Predictions

The predictions from the PNW NHD SPARROW models were analyzed and presented in three ways for this report. First, the models were used to estimate the mean annual incremental TN and TP yield for each of the 232,811 modeled catchments. Incremental yield is equal to the estimated stream load per unit area that is attributable to nutrient sources located exclusively within each incremental catchment, and is a useful tool for comparing the relative intensity of stream load between catchments because it normalizes for contributing area. The median incremental TN and TP yields were then calculated for each of the 22 six-digit hydrologic unit code (HUC6) watersheds (table 1) within the study domain. Second, the models were used to identify the largest local source of TN and TP (that is, the nutrient source contributing the most to the incremental TN and TP yield for each catchment). To simplify the presentation of the results, the nutrient sources were generalized into categories that represented similar activities or processes. The incremental catchments were then grouped together by their largest local source of TN and TP and the median incremental TN and TP yields for each group of catchments were calculated and analyzed for statistical differences in their median values using the Wilcoxan rank sum test with a Simes-Hochberg correction applied to the test values (Simes, 1986; Hochberg, 1988). Third, the models were used to estimate the contribution from each nutrient source to the total TN and TP loads predicted for each reach. Total load was the predicted load contributed from all upstream landscape nutrient sources. The Willamette and Snake Rivers were then used as examples to show how the relative contribution to total load from different nutrient sources varied along two large rivers.

Results

Model Calibration

The TN model included 10 nutrient source terms and 4 delivery terms and the TP model included 9 nutrient source terms and 2 delivery terms (tables 3 and 4). The signs of the delivery terms, rather than the magnitudes, provide information on how they influenced the models. Three delivery terms in the TN model (mean annual precipitation, mean annual solar radiation, and arid land irrigation) had positive coefficients and acted to enhance the delivery of nitrogen from land to water. One delivery term (base flow index) had a negative coefficient and acted to attenuate the delivery of nitrogen from land to water. One delivery term in the TP model (mean annual precipitation) had a positive coefficient and one delivery term (base flow index) had a negative coefficient. Attenuation was not a significant removal mechanism in free-flowing streams or impoundments in either the TN or the TP model.

Based on the R^2 of yield and the RMSE values, the TN model showed better fit with the calibration data set than the TP model (tables 3 and 4), and the NLLS coefficient estimates for both the TN and TP models were generally close in value to the nonparametric bootstrap estimates (meaning that the uncertainty of the NLLS coefficients was low). The exceptions were the coefficients for scrubland and grassland in the TP model and atmospheric deposition and arid land irrigation in the TN model. The negative value for scrubland and grassland in the TP model indicated that there was large uncertainty associated with the coefficient for this source term. Based on an evaluation of the model residuals, the best model fit for TN was in the lower Columbia River basin (LCOL) and the best model fit for TP was in the northern Oregon coastal drainages (NOCR) (appendix B; figs. B1 and B2). The poorest model fit for TN was in the Yakima River basin (with no substantial bias toward over or under prediction) and the poorest model fit for TP was in the Snake River headwaters (due exclusively to under prediction). There were two clusters of under prediction that might have resulted from underestimation of natural nutrient sources. The cluster of under prediction in the Middle Snake-Boise River basin was in an area with documented deposits of nitrate salts (Mansfield and Boardman, 1932),

which were not accounted for in the TN model, and the Snake River headwaters lie in a predominantly forested watershed in a region of extensive phosphate deposits called the Western Phosphate Field (U.S Geological Survey, 2002).

The nitrogen and phosphorus content of livestock manure was represented by two nutrient source terms: confined cattle and grazing livestock. The confined cattle source term represented the manure from cattle associated with a registered dairy or feedlot, and the grazing livestock source term primarily represented the manure from cattle that were not associated with a registered dairy or feedlot, but also included manure from a relatively small number of other non-poultry animals. These two types of manure were included as distinct nutrient sources in the TP model. In contrast, these two nutrient sources were combined and modeled as one source in the TN model because of difficulties experienced during model calibration. Specifically, there was a conflict between the nutrient source terms representing grazing livestock manure and atmospheric deposition; neither source term was significant when both were included in the model along with confined cattle manure. However, when confined cattle and grazing livestock manure were combined and represented by one nutrient source term, the resulting TN model included significant coefficients for both livestock manure (confined and grazing) and atmospheric deposition. Combining the two types of manure in the TN model was justified because the coefficient values of the confined cattle and grazing livestock source terms were similar when modeled separately, which indicated no substantial difference in availability.

The decision to include source terms representing nutrients from nonpoint urban sources and springs and power returns was not based solely on the statistical results from the calibrations. The nitrogen and phosphorus loading from nonpoint urban sources was represented by the area of developed land rather than by an alternative surrogate equal to non-farm fertilizer use (appendix A). Either one of these nutrient source terms was significant in the TN and TP models, but developed land was selected because it accounted for most nonpoint sources of urban[1] nutrients, including fertilizer use, leaking sewer lines, animal manure, as well as other sources, whereas non-farm fertilizer represented only one source. The load from springs and power returns was a significant nutrient source term in the TN model but not in the TP model (p-value = 0.1408). This nutrient source term was retained in the TP model, however, because it was an important local source of phosphorus for some stream reaches. This determination was based on the large under predictions that were observed at calibration stations located downstream of springs and power returns when they were not included in the TP model calibration.

---

[1] The exception was nitrogen leaching from septic tanks, which was modeled as a separate source in areas that were not served by municipal sewer lines in 2002.

Table 3. Model Statistics for the total nitrogen National Hydrography Dataset SPARROW model developed for the United States Pacific Northwest.

[The p-values for the source and aquatic loss variables are based on a one-sided t-test; the p-values for the land-to-water delivery variables are based on a two-sided t-test. **Abbreviations:** NLLS, non-linear least squares; R^2, coefficient of determination; RMSE, root mean squared error; cm, centimeter; kg/yr, kilogram per year; kg/km²-yr, kilogram per square kilometer per year; kg, kilogram; km²; square kilometer; m², square meter; mm, millimeter; m²-day, square meter per day; <, less than; –, not applicable]

Parameter	Model coefficient units	NLLS estimate of model coefficient	Standard error of the model coefficient	Probability level (p-value)	90 percent confidence interval for the model coefficient		Nonparametric bootstrap estimate of model coefficient
					Lower	Upper	
Sources							
Forestland (west)[1] (km²)	kg/km²-yr	25.5	9.8	0.0104	3.7	41.6	26.8
Forestland (east)[2] (km²)	kg/km²-yr	71.6	10.7	< 0.0001	49.5	113.5	76.2
Point sources[3] (kg/yr)	dimensionless	1.03	0.38	0.0075	0.40	1.48	1.04
Developed land[4] (km²)	kg/km²-yr	337	69	< 0.0001	253	409	334
Nonsewered population[5] (number)	kg/person-yr	1.17	0.65	0.0735	0.24	1.75	1.08
Atmospheric deposition[6] (kg/yr)	dimensionless	0.026	0.012	0.0322	-0.082	0.042	0.007
Farm fertilizer[7] (kg/yr)	dimensionless	0.069	0.012	< 0.0001	0.046	0.097	0.072
Livestock manure[8] (kg/yr)	dimensionless	0.072	0.030	0.0172	0.025	0.143	0.083
Red alder trees[9] (m²)	kg/m²-yr	0.28	0.10	0.0054	0.08	0.41	0.26
Springs and power returns[10] (kg/yr)	dimensionless	0.96	0.35	0.0069	0.43	1.20	0.91
Land-to-water delivery							
Precipitation[11] (mm)	–	1.29	0.14	< 0.0001	1.11	1.78	1.38
Solar radiation[12] (watt-hours/m²-day)	–	7.85	1.83	< 0.0001	-0.68	11.03	6.59
Arid land irrigation[13] (percent)	–	0.015	0.005	0.0012	0.006	0.031	0.035
Base flow index[14] (percent)	–	-0.022	0.005	0.0001	-0.031	-0.009	-0.021
Aquatic loss							
Free-flowing streams	–	–	–	–	–	–	–
Impoundments	–	–	–	–	–	–	–
Model diagnostics							
R^2 of yield	–	0.908	–	–	–	–	–
RMSE	–	0.398	–	–	–	–	–
Number of observations	–	172	–	–	–	–	–

[1] Area of forest land, west side of Cascade Range, 2001.

[2] Area of forest land, east side of Cascade Range, 2001.

[3] Surface-water discharges from permitted wastewater facilities, 2002.

[4] Area of developed land, 2001.

[5] Population without sewer service, 2001.

[6] Wet and dry atmospheric deposition of oxidized and reduce nitrogen, 2002.

[7] Commercial fertilizer applied to agricultural land, 2002.

[8] Manure from cattle in dairies and feedlots and grazing livestock, 2002.

[9] Basal area of red alder trees, circa 2002.

[10] Surface-water discharge from springs and power returns, 2002.

[11] Natural log of mean annual precipitation (1971–2000).

[12] Natural log of mean annual solar radiation (1991–2005); interaction limited to the forest land source terms.

[13] Percentage of catchment containing irrigated agriculture; interaction limited to the farm fertilizer source term.

[14] Mean percentage of streamflow due to base flow.

Table 4. Model statistics for the total phosphorus National Hydography SPARROW model developed for the United States Pacific Northwest.

[The p-values for the source and aquatic loss variables are based on a one-sided t-test; the p-values for the land-to-water delivery variables are based on a two-sided t-test. **Abbreviations:** NHD, National Hydrography Dataset; NLLS, non-linear least squares; RMSE, root mean squared error; R^2, coefficient of determination; cm, centimeter; kg/km^2-yr, kilogram per square kilometer per year; kg/yr, kilogram per year; km^2, square kilometer; mm, millimeter; <, less than; –, not applicable]

Parameter	Model coefficient units	NLLS estimate of model coefficient	Standard error of the model coefficient	Probability level (p-value)	90 percent confidence interval for the model coefficient		Nonparametric bootstrap estimate of model coefficient
					Lower	Upper	
Sources							
Forestland (west)[1] (km^2)	kg/km^2-yr	9.23	2.87	0.0015	5.47	14.08	10.15
Forestland (east)[2] (km^2)	kg/km^2-yr	6.81	1.42	< 0.0001	4.29	12.24	8.43
Scrub and grass land[3] (km^2)	kg/km^2-yr	0.54	0.32	0.0898	-10.66	0.77	-3.28
Point sources[4] (kg/yr)	dimensionless	1.20	0.33	0.0003	0.63	1.55	1.17
Developed land[5] (km^2)	kg/km^2-yr	24.3	5.5	< 0.0001	19.7	29.9	24.6
Farm fertilizer[6] (kg/yr)	dimensionless	0.015	0.007	0.0271	0.002	0.025	0.014
Confined cattle[7] (kg/yr)	dimensionless	0.035	0.014	0.0100	0.019	0.060	0.038
Grazing livestock[8] (kg/yr)	dimensionless	0.124	0.022	< 0.0001	0.094	0.195	0.143
Springs and power returns[9] (kg/yr)	dimensionless	1.67	1.13	0.1408	1.17	2.53	1.76
Land-to-water delivery							
Precipitation[10] (mm)	–	1.26	0.17	< 0.0001	0.86	1.52	1.20
Base flow index[11] (percent)	–	-0.012	0.006	0.0340	-0.024	-0.001	-0.012
Aquatic loss							
Free-flowing streams	–	–	–	–	–	–	–
Impoundments	–	–	–	–	–	–	–
Model diagnostics							
R^2 of yield	–	0.810	–	–	–	–	–
RMSE	–	0.549	–	–	–	–	–
Number of observations	–	220	–	–	–	–	–

[1] Area of forest land, west side of Cascade Range, 2001.

[2] Area of forest land, east side of Cascade Range, 2001.

[3] Area of scrub and grass land, 2001.

[4] Surface-water discharges from permitted wastewater facilities, 2002.

[5] Area of developed land, 2001.

[6] Commercial fertilizer applied to agricultural land, 2002.

[7] Manure from cattle in dairies and feedlots, 2002.

[8] Manure from grazing livestock, 2002.

[9] Surface-water discharge from springs and power returns, 2002.

[10] Natural log of mean annual precipitation, 1971–2000.

[11] Mean percentage of streamflow due to base flow.

Model Predictions

The incremental yields of TN and TP were much greater on the western side of the Cascade Range compared to the eastern side (table 5; figs. B3 and B4). The largest median incremental yields of TN and TP were predicted for the northern Oregon coast (NOCR) and the Washington coast (WACR), respectively. The high yields of TN and TP in these watersheds were directly related to the large amount of precipitation. The incremental yields of TN and TP in other areas west of the Cascade Range (PUGT, LCOL, WILL, and SOCR) were less than in WACR and NOCR, but still greater than areas east of the Cascade Range. In the east side watersheds the greatest median incremental TN yields were predicted in the Clearwater (CLRW) and Middle Columbia River (MCOL) basins and the greatest median incremental TP yields were predicted in the Clearwater and Spokane River (SPOK) basins. The large median TN and TP yields in the Clearwater River basin were due to high precipitation compared to the other east side HUC6 watersheds. The large median TN yield in the Middle Columbia River basin was due to a combination of high average input from farm fertilizer and high average precipitation compared to the other east side HUC6 watersheds and the large median TP yield in the Spokane River basin was due to high levels of urbanization in this watershed compared to other east side HUC6 watersheds.

Table 5.   Median incremental yields of total nitrogen and total phosphorus for six-digit Hydrologic Unit Code (HUC6) watersheds in the United States Pacific Northwest (2002 conditions).

[HUC6 locations are shown in figures 1–3. Values are based on the results obtained from then National Hydrography Dataset (NHD) total nitrogen and total phosphorus SPARROW models developed for the United States Pacific Northwest. **Incremental yield:** is equal to the local load divided by the NHD catchment area, where the local load is the load generated exclusively within an NHD catchment. **Abbreviation:** kg/ha-yr, kilogram per hectare per year]

Region	HUC6 abbreviation	Total nitrogen		Total phosphorus	
		Median incremental yield (kg/ha-yr)	Rank	Median incremental yield (kg/ha-yr)	Rank
Pacific Coast	WACR	6.25	2	0.745	1
	NOCR	10.87	1	0.630	2
	SOCR	2.22	6	0.403	6
West Side Basins	PUGT	3.88	5	0.540	3
	LCOL	4.05	4	0.508	5
	WILL	4.55	3	0.517	4
Columbia River Basin	KOOT	0.51	16	0.098	11
	PDOR	0.63	13	0.128	10
	SPOK	0.87	11	0.140	8
	UCOL	0.42	17	0.063	19
	YAKI	0.93	10	0.070	13
	JDAY	0.41	18	0.066	16
	DESC	0.28	20	0.068	14
	MCOL	1.10	8	0.066	17
Snake River Basin	SNKH	0.57	14	0.065	18
	USNK	0.33	19	0.050	20
	MSBS	0.10	21	0.026	22
	MSPW	0.55	15	0.130	9
	SALM	0.69	12	0.068	15
	CLRW	1.13	7	0.141	7
	LSNK	0.94	9	0.097	12
Oregon Closed Basins	ORCB	0.10	22	0.040	21

The largest local sources of TN and TP load are shown in figures 2 and 3 and summarized in table 6. The figures do not show catchments where the largest local source was the load at a boundary reach or the load from springs and power returns because these represented less than 1 percent of the total number of catchments. The largest local source of TN in more than 50 percent of the catchments was from forestland or alder trees and the largest local source of TP in almost 90 percent of the catchments was from livestock manure (primarily grazing livestock) and geologic sources. The highest median TN and TP yields were predicted for catchments where urban sources were the largest local source, the lowest median TN yields were predicted for catchments where atmospheric deposition was the largest local source, and the lowest median TP yields were predicted for catchments where farm fertilizer was the largest local source (table 6). There were significant ($\alpha < 0.05$) differences between each of the median incremental TN yields and each of the median incremental TP yields shown in table 6.

On average, the largest contributor to the total TN load in PNW streams was forestland, which was responsible for at least 63 percent of the TN load in one-half of the reaches (table 7). On average, the largest contributors to total TP load in PNW streams were geologic phosphorus (which was represented by forestland, scrubland, and grassland) and livestock manure (primarily grazing livestock). In one-half of the reaches these two nutrient sources were responsible for at least 58 and 30 percent, respectively, of the TP load. Although on average diffuse nutrient sources (both natural

and anthropogenic) were responsible for most of the total TN and TP load in PNW streams, concentrated anthropogenic nutrient sources contributed much of the total load in some of the large rivers. Two examples were the Willamette River in western Oregon (fig. 4) and the Snake River, which flows through southern Idaho, northeastern Oregon, and southeastern Washington (fig. 5). Both rivers drain watersheds containing a mix of agricultural, urban, and undeveloped land. Forestland and farm fertilizer were the largest contributors to TN load in the Willamette River upstream of Eugene and Springfield and farm fertilizer was generally the largest contributor to TN load downstream of this point. The exception was the 45 km of river immediately downstream of Eugene and Springfield, where the inputs of nitrogen from urban sources (primarily wastewater treatment plants) resulted in point sources being the largest contributor to TN load. Geologic phosphorus was the largest contributor to TP load in the Snake River between its headwaters and Idaho Falls, whereas point sources and agricultural nutrient sources (farm fertilizer and livestock manure) were generally the largest contributors downstream of this point. The livestock manure generated along this part of the Snake River was mostly from cattle in dairies and feedlots. Almost all of the contribution from urban nutrient sources upstream of the Boise River was from fish farms and hatcheries[2], whereas downstream, a large percentage was from wastewater treatments plants, especially those that discharged to the Boise River. The three-fold increase in TP load between Twin Falls and King Hill was mostly due to phosphorus input from springs and the large number of fish farms and hatcheries located along this reach. The contribution from urban nutrient sources (almost exclusively fish farms and hatcheries) was as high as 50 percent of the total TP load in this segment of the Snake River. The three-fold increase in TP load downstream of the Boise River was mostly due to phosphorus input from this large tributary.

---

[2] Although fish farms and hatcheries are not typically in urban areas they were grouped with urban sources in this study because their effects could not be distinguished from urban point sources during model calibration.

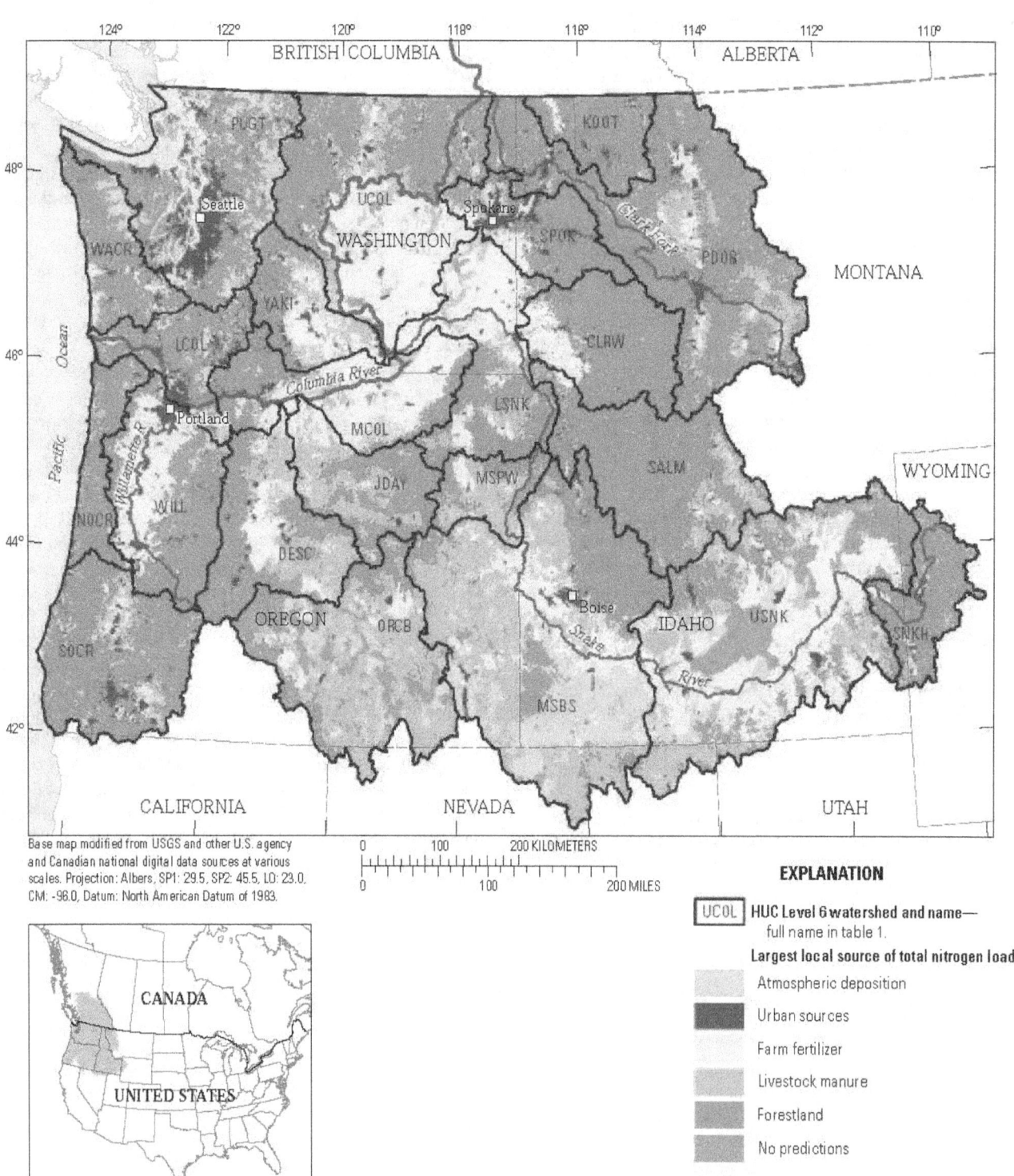

Figure 2. Largest local sources of total nitrogen for National Hydrography Dataset catchments in the United States Pacific Northwest (2002 conditions).

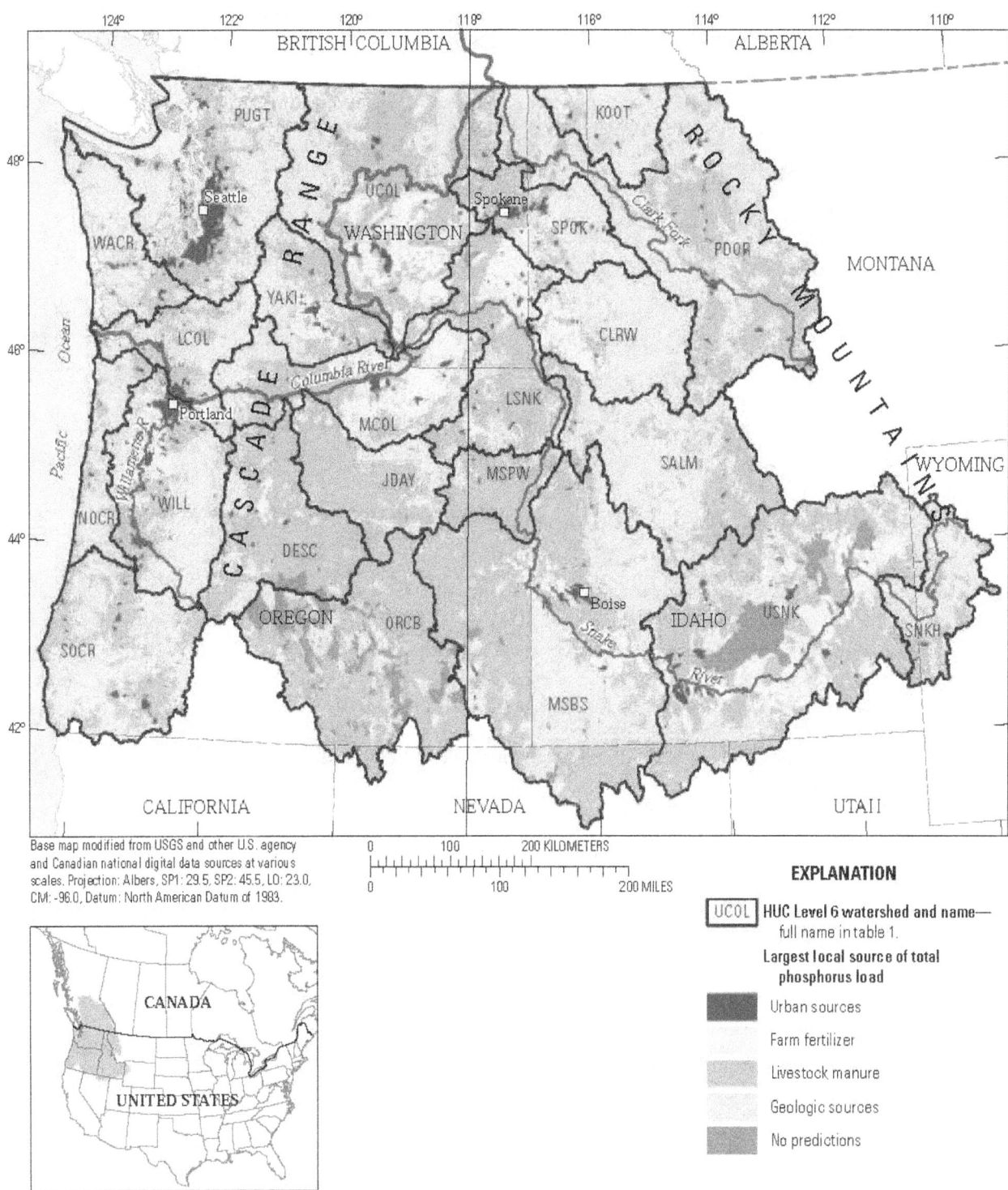

Figure 3.   Largest local sources of total phosphorus for National Hydrography Dataset catchments in the United States Pacific Northwest (2002 conditions).

Table 6. Largest local sources of total nitrogen and total phosphorus and corresponding median incremental yields for National Hydrography Dataset catchments in the United States Pacific Northwest (2002 conditions).

[**Catchments:** Percentage of catchments where the source category was the largest contributor to the local load within catchments where local load is equal to the load generated exclusively within an National Hydrography Dataset (NHD) catchment. **Median incremental yield:** for catchments where the source was the largest contributor to the local load. Incremental yield is equal to the local load divided by the incremental catchment area. **Abbreviations:** kg/ha-yr, kilogram per hectare per year; –, not applicable because source was not a source term used in this model or was the largest source in a small percentage of catchments; <, less than]

Largest local source category	Total nitrogen		Total phosphorus	
	Catchments (percent)	Median incremental yield (kg/ha-yr)	Catchments (percent)	Median incremental yield (kg/ha-yr)
Atmospheric deposition	10.5	0.037	–	–
Urban sources[1]	2.92	2.47	1.40	0.42
Farm fertilizer	15.63	1.96	8.43	0.07
Livestock manure[2]	16.5	0.14	39.1	0.09
Geologic sources[3]	–	–	49.5	0.11
Forestland[4]	53.7	0.91	See footnote [6]	
Springs and power returns	< 1.0	–	< 1.0	–
Unidentified[5]	< 1.0	–	1.52	–

[1] Contribution from developed land, point sources, and nitrogen leaching from septic tanks.

[2] Contribution from all livestock manure.

[3] Contribution from geologic phosphorus exported from forest, scrub, and grass land.

[4] Contribution from asymbiotic nitrogen fixation in forests and symbiotic nitrogen fixation by red alder trees.

[5] Catchments where no model predictions were made because they were internal sinks or, in the case of total phosphorus, there were no modeled nutrient sources.

[6] Phosphorus export from forest land is included in the geologic phosphorus source category.

Table 7. Summary statistics for source shares of total nitrogen and total phosphorus load for National Hydrography Dataset stream reaches in the United States Pacific Northwest (2002 conditions).

[**Total nitrogen and phosphorus load, source shares:** The contribution from each source as a percentage of the total load due to all upstream landscape nutrient loading. **Abbreviations:** NHD, National Hydrography Dataset; –, not applicable because this was not a source used in the model]

Source category	Total nitrogen load (source shares)						Total phosphorus load (source shares)					
	Mean	Percentiles					Mean	Percentiles				
		10th	25th	50th	75th	90th		10th	25th	50th	75th	90th
Atmospheric deposition	17.3	3.38	6.51	11.5	19.6	38.0	–	–	–	–	–	–
Urban sources[1]	2.78	0.00	0.00	0.0	0.5	6.3	1.73	0.00	0.00	0.00	0.02	1.26
Farm fertilizer	12.4	0.00	0.00	0.0	3.11	65.6	7.26	0.00	0.00	0.00	0.62	19.1
Livestock manure[2]	17.1	0.45	2.31	7.9	21.4	53.0	37.6	1.66	10.4	30.2	62.4	87.4
Geologic sources[3]	–	–	–	–	–	–	53.1	3.88	17.8	58.0	85.6	96.8
Forestland[4]	49.9	0.00	7.89	62.6	80.5	88.4	See footnote[5]					
Springs and power returns	0.18	0.00	0.00	0.00	0.00	0.00	0.12	0.00	0.00	0.00	0.00	0.00

[1] Contribution from developed land, point sources, and nitrogen leaching from septic tanks.

[2] Contribution from all livestock manure.

[3] Contribution from geologic phosphorus exported from forest, scrub, and grass land.

[4] Contribution from asymbiotic nitrogen fixation in forests and symbiotic nitrogen fixation by red alder trees.

[5] Phosphorus export from forest land is included in the geologic phosphorus source category.

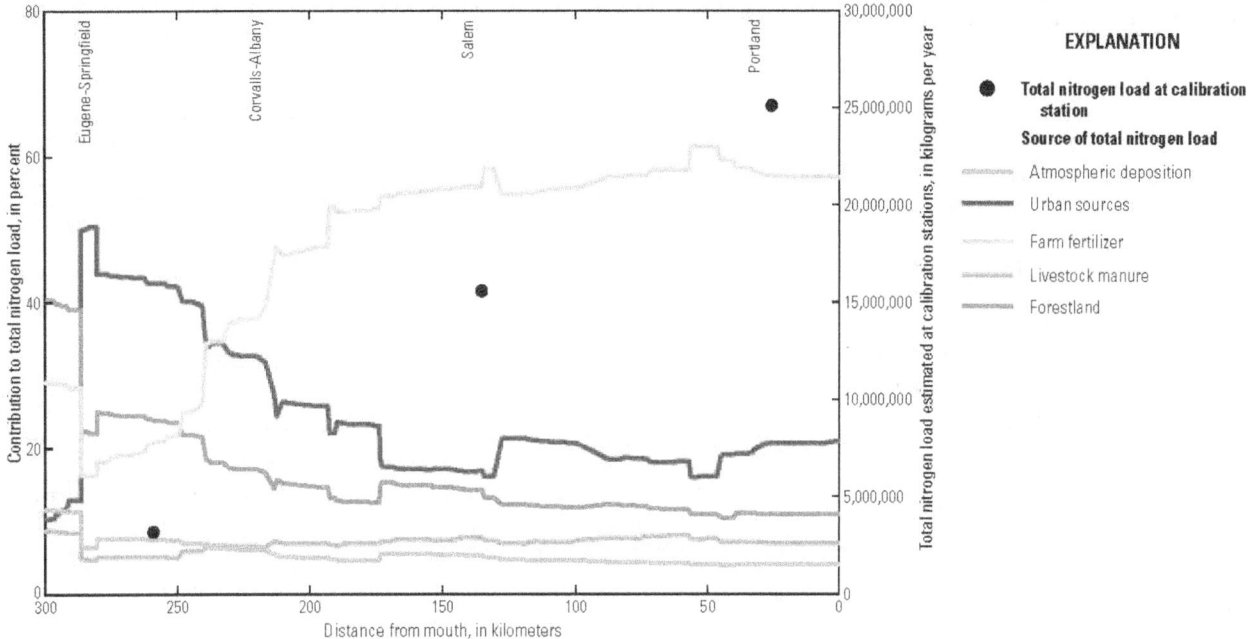

Figure 4.   Results for the Willamette River, Oregon, from the total nitrogen National Hydrography Dataset SPARROW model developed for the United States Pacific Northwest (2002 conditions). Urban sources include developed land, discharge from wastewater-treatment facilities, and leaching from septic tanks; livestock manure is generated by confined and grazing livestock.

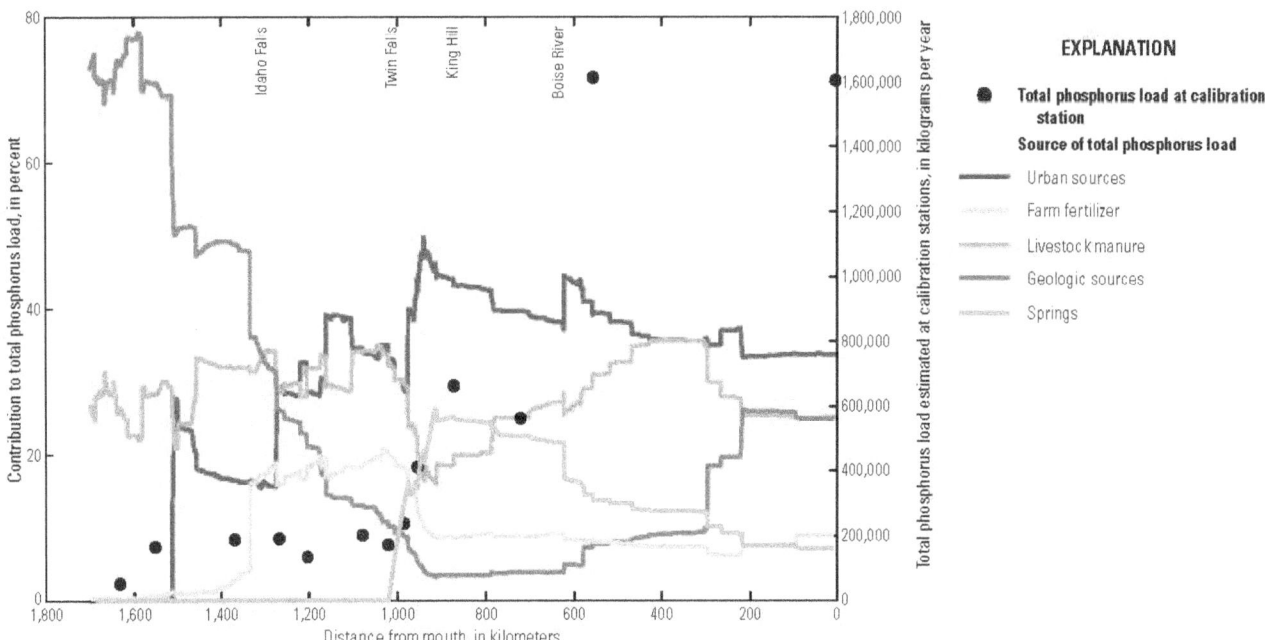

Figure 5.   Results for the Snake River, Idaho, Oregon, and Washington, from the total phosphorus National Hydrography Dataset SPARROW model developed for the United States Pacific Northwest (2002 conditions). Urban sources include developed land and discharge from wastewater-treatment facilities and fish hatcheries; geologic sources include phosphorus runoff from forestland, scrubland, and grassland; livestock manure is generated by confined and grazing livestock.

Discussion of Results

The calibrations of the PNW NHD SPARROW models resulted in a better overall fit with the calibration load data compared to the RF1 models (table 8). These improvements in model fit were likely due to a combination of factors: (1) the improved resolution of the NHD network compared to the RF1 network, (2) differences in the set of nutrient loads used to calibrate the models, and (3) refinements in the estimates of anthropogenic nutrient sources. The better fit of the NHD models compared to the RF1 models resulted in less uncertainty in the predictions of nutrient loads, incremental nutrient yields, and nutrient source shares. Model uncertainty associated with the PNW RF1 models, which also applies to the PNW NHD models, is discussed in Wise and Johnson (2011).

The coefficients for the nutrient source terms (tables 3 and 4) provide some insight into why certain results were obtained. The coefficients for the nutrient source terms representing land cover types (expressed as kilograms per square kilometers per year [(kg/km^2)-yr]) were analogous (but not directly comparable) to yield values or export coefficients, whereas the coefficients for the nutrient source terms representing mass loading (expressed as kilograms per year [kg/yr]) were an indication of the availability for delivery to streams. In both the TN and TP models, developed land had the largest coefficient of any land cover source term, which reflected the large number of nonpoint nutrient sources in urban areas and the relative ease with which they are transported to streams. The highest incremental yields were predicted for catchments dominated by urban nutrient sources because of the large coefficients for developed land and the fact that nutrients from point sources were discharged directly to streams. In both models the land cover terms representing natural nutrient sources had much smaller coefficients than developed land, which explained why catchments where the largest share of the local load was from these land cover types had relatively low incremental yields.

The coefficients for the farm fertilizer source terms indicated that a greater proportion of nitrogen was delivered to streams compared to phosphorus and this explained why catchments where the largest share of the local load was from farm fertilizer had relatively high incremental TN yields but the lowest incremental TP yields. These results are consistent with the physical and chemical processes affecting the mobility of nitrogen and phosphorus. Although the coefficients for the livestock manure source terms in both models were greater than those for the farm fertilizer source terms, catchments where the largest share of the local load was from livestock manure had relatively low incremental yields. This was because most of the manure was from grazing livestock, which represents a more diffuse source of nutrients compared to farm fertilizer. The coefficients for the source term representing point sources and the source term representing springs and power returns in the TN model were about 1, but the coefficients for these nutrient source terms in

Table 8. Comparison of fit statistics for the National Hydrography Dataset and RF1 SPARROW models developed for the United States Pacific Northwest.

[**Abbreviations:** NHD, National Hydrography Dataset; RF1, River Reach File 1; R^2, coefficient of determination; RMSE, root mean squared error]

Fit Statistic	Total nitrogen		Total phosphorus	
	NHD model	RF1 model	NHD model	RF1 model
R^2 of yield	0.908	0.759	0.810	0.712
RMSE	0.398	0.640	0.549	0.693

the TP model were 1.20 and 1.67, respectively. Because these coefficients acted as scaling factors for the estimated load, the model results indicated that our estimates of TP load from point sources and from springs and power returns were likely too low.

There were potential advantages to using a geologic phosphorus source term that was based on a phosphorus index instead of the land cover source terms that were used in the NHD TP model. This was the approach that was used in RF1 SPARROW TP model (Wieczorek and Lamotte, 2013; Wise and Johnson, 2011). One potential advantage was that the phosphorus index provided a spatially continuous distribution of geologic phosphorus. The source terms for land cover did not account for geologic phosphorus released from land cover types other than forestland, scrubland, and grassland. Geologic phosphorus released from agricultural and developed land was not a significant nutrient source in the TP model because it was negligible compared to the contributions from anthropogenic activities (farm fertilizer, confined cattle, and developed land). Another potential advantage was that the phosphorus index accounted for regional differences in the phosphorus content of geologic material whereas the land cover source terms did not. There was no evidence, however, that using the phosphorus index provided a better model fit in areas with relatively high concentrations of geologic phosphorus. The under predictions at the calibration stations located in the Snake River headwaters (which are in the Western Phosphate Field), for example, were greater when the phosphorus index was used in place of the land cover source terms.

The better fit for the TP model using the source terms for land cover compared to the phosphorus index indicated that weathering processes were a more important control on the availability of natural phosphorus than the phosphorus content of the rocks themselves. Additionally, the coefficient for the west-side forest source term being the largest of the three geologic source terms indicated that the geologic phosphorus in this land cover type was more readily delivered to streams compared to the geologic phosphorus in the other land cover types (east side forests and scrubland/grassland). These results were most likely due to differences between the amount of precipitation that falls on the west and east sides of the Cascade Range. The mean annual precipitation in predominantly (greater than 50 percent) forestland catchments on the west side of the Cascade Range (194 cm) was greater than the mean annual precipitation in predominantly forestland catchments on the east side (84 cm) and predominantly scrubland and grassland catchments (43 cm; almost all located on the east side). Soil erosion rates in forested catchments on the west side of the Cascade Range are greater than soil erosion rates in forested catchments on the east side (Elliot, 2006) and, presumably, also greater than soil erosion rates in scrubland and grassland. As a result, the export of geologic phosphorus from west-side forestland was expected to be greater than the export of geologic phosphorus from east-side forestland, scrubland, and grassland. The relatively small coefficient for the scrubland and grassland source term (compared to the forestland terms) might have reflected slower weathering processes in arid and generally low slope environments.

The calibration results for the delivery terms in the TN and TP models were consistent with our assumptions about the processes that occur within watersheds. The delivery of nitrogen and phosphorus to streams in catchments receiving high precipitation was larger than in similar catchments receiving less precipitation. Similarly, in arid areas agricultural catchments having a large percentage of irrigated land contributed more nitrogen to nearby streams than similar catchments having less irrigated land. Catchments having a large baseflow index tended to contribute less nitrogen than catchments having a small baseflow index, presumably because of denitrification in soil and groundwater. The contribution of phosphorus also was less from catchments having a large baseflow index compared to catchments having a small baseflow index. This most likely resulted from less overland flow in catchments having a large baseflow index, meaning that less sediment-bound phosphorus was available for transport. The use of solar radiation as a delivery term is unique to the PNW NHD SPARROW model for TN. The positive relation between solar radiation and forest productivity could explain the significance of this delivery term. Forested catchments that receive more solar radiation have greater productivity rates and, presumably greater nitrogen uptake from the soil than forested watersheds that receive less solar radiation. This could leave less nitrogen available for permanent removal through soil denitrification.

The results obtained for nutrient loss in free-flowing streams and impoundments in the NHD models were not typical of recent SPARROW nutrient applications for large hydrologic regions (see Hoos and McMahon, 2009; Brown and others, 2011; Garcia and others, 2011; Moore and others, 2011; Rebich and others, 2011; and Robertson and Saad, 2011) but were consistent with the PNW RF1 models. All the other regional SPARROW applications identified TN loss in free-flowing streams as a significant removal process, whereas the PNW NHD TN model did not. The result could have reflected the importance of fixation in the cycling of nitrogen in PNW streams, meaning that a substantial amount of the nitrogen lost through denitrification was replaced with nitrogen fixed from the atmosphere. All of the other regional SPARROW applications also identified TP loss in impoundments as a significant removal process whereas the PNW NHD TP did not. This result indicates that settling within impoundments was not an important mechanism for phosphorus removal in PNW surface waters. This could have been because a large proportion of the phosphorus was in the dissolved form (and was not settling out before flowing out of the impoundments) or because the settling of phosphorus in impoundments was balanced by phosphorus dissolution and resuspension.

Conclusions

Results from the TN and TP models showed that anthropogenic nutrient sources were responsible for a substantial part of the nutrient load in some reaches. The models also showed, however, that the largest average contribution to TN and TP load in PNW streams was from natural nutrient sources and that most of the contribution from anthropogenic nutrient sources was from diffuse rather than point sources. This means that regulatory actions taken to manage nutrient loads should consider the importance of natural sources and diffuse anthropogenic sources such as atmospheric deposition, farm fertilizer, and grazing livestock as well as point sources. Finally, although urban sources were the largest local source of TN and TP in only a small percentage of the incremental catchments, the concentration of these catchments around Portland and the Puget Sound indicated that anthropogenic urban sources in these highly developed areas were responsible for a substantial amount of the nutrients in surface water.

The findings from this study could help complement research and inform water-quality management in the PNW. Two examples were presented that showed how the SPARROW model can be used to assess nutrient conditions in large rivers that drain watersheds with a mix of land cover. The Willamette Ecosystem Services Project being developed by the U.S. Environmental Protection Agency will examine how land use and human activities affect the distribution of reactive nitrogen in the Willamette Basin (Compton and others, 2009). The results from the SPARROW TN model described in this report could complement the Willamette Ecosystem Services Project by showing the distribution of nitrogen in surface waters within the Willamette Basin. The approach used to analyze TN loads and sources in the Willamette River also could be used as a template for estimating the contribution from different sources to the nitrogen load in tributary watersheds. The States of Idaho and Oregon have designated most of the Snake River below Twin Falls as water-quality impaired and the resulting Total Maximum Daily Loads (TMDLs) are focused on reducing phosphorus inputs to this stretch of the river. The results of this study provided a complete description of the TP load throughout the Snake River Basin and the relative contribution from all major phosphorus sources to that load, information that should be useful when TMDLs are developed or refined in the future. In addition to the model results, researchers and water-quality managers might find value in the input data sets used in this study. These data sets are the most comprehensive collection of surface-water nutrient measurements and landscape nutrient loading estimates compiled to date for the PNW.

Summary

This regional assessment of surface-water nutrients used the National Hydrography Dataset Plus (NHDPlus) stream flow network and updated calibration data to develop refined SPARROW models for TN and TP for the PNW. These new models improve upon and supersede the earlier RF1-based models for the PNW. The PNW NHD SPARROW models were used to estimate mean annual nutrient conditions for 2002 and to identify the relative contribution of different sources to instream nutrient loads, both at established water-quality monitoring stations and in stream reaches where little or no monitoring data were available. The input data sets for the NHD models and the model results that resulted from them should be useful to researchers and water-quality managers when performing local nutrient assessments. The model results were used to show how the relative contributions from natural and anthropogenic nutrient sources (point sources and diffuse sources) change along two large rivers between their headwaters and mouths. Future SPARROW nutrient modeling for the PNW should be based on a more recent time period and could reflect mean seasonal as well as mean annual conditions. Additionally, sub-regional SPARROW models could be developed to help inform water-quality management decisions in specific areas of the PNW.

Acknowledgments

We thank Michael Wieczorek (U.S. Geological Survey) for compiling many of the catchment attribute data sets used in this study and Esther Duggan-Pischel and Tana Haluska (U.S. Geological Survey) for helping to compile the catchment attribute data sets that were developed specifically for the Pacific Northwest. We thank the agencies that provided the water-quality data used in this study and Kenneth Skach (U.S. Geological Survey) for helping to compile those data.

References Cited

Alexander, R.B., Smith, R.A., Schwarz, G.E., Boyer, E.W., Nolan, J.V., and Brakebill, J.W., 2008, Differences in phosphorus and nitrogen delivery to the Gulf of Mexico from the Mississippi River Basin: Environmental Science and Technology, v. 42, no. 3, p. 822–830.

Brakebill, J.W., Wolock, D.M., and Terziotti, S.E., 2011, Digital hydrologic networks supporting applications related to spatially referenced regression modeling: Journal of the American Water Resources Association, v. 47, no. 5, p. 1752–1688.

Brown, J.B., Sprague, L.A., and Dupree, J.A., 2011, Nutrient sources and transport in the Missouri River Basin, with emphasis on the effects of irrigation and reservoirs: Journal of the American Water Resources Association, v. 47, no. 5, p. 1034–1060.

Compton, J.E., Dennis, R.L, Walker, H.A., Milstead, W.B., Jordan , S.J., Hill, B.H., Fritz, K.M., Devereux, Richard, Johnson, B.R., Beaulieu, J.J., Latimer, J.S., Lynch, J.A., Waite R.G., and Davis C.C., 2009, Linking ecosystem services and nitrogen—Science to improve management of nitrogen in air, land and water: Corvallis, Oreg., U.S. Environmental Protection Agency, Office of Research and Development, National Health and Environmental Effects Research Laboratory–Western Ecology Division, 82 p.

Elliot, W.J., 2006, The roles of natural and human disturbances in forest soil erosion, in Owens, P.N., and Collins, A.J., eds., Soil erosion and sediment redistribution in river catchments: Oxfordshire, United Kingdom, Centre for Agricultural Bioscience International, 328 p.

Garcia, A.M., Hoos, A.B., and Terziotti, Sylvia, 2011, A regional modeling framework of phosphorus sources and transport in streams of the southeastern United States: Journal of the American Water Resources Association, v. 47, no. 5, p. 991–1010.

Hochberg, Yossi, 1988. A sharper Bonferroni procedure for multiple tests of significance: Biometrika, v. 75, p. 800–802.

Homer, Collin, Huang, Chengquan, Yang, Limin, Wylie, Bruce, and Coan, Michael, 2004, Development of a 2001 national land cover database for the United States: Photogrammetric Engineering and Remote Sensing, v. 70, p. 829–840.

Hoos, A.B., and McMahon, Gerard, 2009, Spatial analysis of instream nitrogen loads and factors controlling nitrogen delivery to streams in the southeastern United States using Spatially Referenced Regressions on Watershed Attributes (SPARROW) and regional classification frameworks: Hydrological Processes, v. 22, no. 16, p. 2275–2294.

Horizon Systems, 2013, NHDPlusV2Data: Horizon Systems database, accessed March 18, 2013, at http://www.horizon-systems.com/nhdplus/.

Mansfield, G.R., and Boardman, Leona, 1932, Nitrate deposits of the United States: U.S. Geological Survey Bulletin 838, 107 p.

Moore, R.B., Johnston, C.M., Smith, R.A., and Milstead, Bryan, 2011, Source and delivery of nutrients to receiving waters in the northeastern and mid-Atlantic regions of the United States: Journal of the American Water Resources Association, v. 47, no. 5, p. 965–990.

National Research Council, 2001, Assessing the TMDL Approach to Water Quality Management: Washington D.C., National Academy Press, 109 p.

Preston, S.D., Alexander, R.B., Woodside, M.D., and Hamilton, P.A., 2009, SPARROW modeling—Enhancing understanding of the Nation's water quality: U.S. Geological Survey Fact Sheet 2009-3019, 6 p.

Rebich, R.A., Houston, N.A., Mize, S.V., Pearson, D.K., Ging, P.B., and Hornig, C.E, 2011, Sources and delivery of nutrients to the northwestern Gulf of Mexico from streams in the south-central United States: Journal of the American Water Resources Association, v. 47, no. 5, p. 1061–1086.

Robertson, D.M., and Saad, D.A., 2011, Nutrient inputs to the Laurentian Great Lakes by source and watershed estimated using SPARROW watershed models: Journal of the American Water Resources Association, v. 47, no. 5, p. 1011–1033.

Saad, D.A., Schwarz, G.E., Robertson, D.M., and Booth, N.L., 2011, A multi-agency nutrient dataset used to estimate loads, improve monitoring design, and calibrate regional nutrient SPARROW models: Journal of the American Water Resources Association, v. 47, no. 5, p. 933–949.

Schwarz, G.E., Hoos, A.B., Alexander, R.B., and Smith, R.A., 2006, The SPARROW surface water-quality model—Theory, applications and user documentation: U.S. Geological Survey Techniques and Methods, book 6, chap. B3, 248 p. and CD-ROM.

Seaber, P.R., Kapinos, F.P., and Knapp, G.L., 1987, Hydrologic unit maps: U.S. Geological Survey Water-Supply Paper 2294, 63 p.

Simes, R.J., 1988, An improved Bonferroni procedure for multiple tests of significance: Biometrika, v. 73, p. 750–754.

Simley, J.D., and Carswell, W.J. Jr., 2009, The National Map—Hydrography: U.S. Geological Survey Fact Sheet 2009-3054, 4 p. (Also available at http://pubs.usgs.gov/fs/2009/3054/.)

Smith, R.A., Schwarz, G.E., and Alexander, A.B., 1997, Regional interpretation of water-quality monitoring data, Water Resources Research, v. 33, no. 12, p. 2781–2798.

U.S. Geological Survey, 2002, Western phosphate field, U.S.A.—Science in support of land management: U.S. Geological Survey Fact Sheet 100-02, 2 p. (Also available at http://pubs.usgs.gov/fs/fs-100-02/.)

U.S. Geological Survey, 2012, National Water Information System (NWISWeb): U.S. Geological Survey database, accessed October 6, 2012, at http://waterdata.usgs.gov/nwis.

Wieczorek, M.E., and Lamotte, A.E., 2013, Attributes for MRB_E2RF1 catchments by major river basins in the conterminous United States: U.S. Geological Survey Digital Data Series DS 491, accessed March 18, 2013 at http://water.usgs.gov/nawqa/modeling/rf1attributes.html.

Wise, D.R., and Johnson, H.M., 2011, Surface-water nutrient conditions and sources in the United States Pacific Northwest: Journal of the American Water Resources Association, v. 47, no. 5, p. 1110–1135.

Appendix A.   Compilation of Model Input Data

This appendix contains brief descriptions of how the data sets representing nutrient sources and land-to-water-delivery variables were compiled and how these data sets differed from those used in the RF1 SPARROW models. Detailed descriptions of the methods used to compile these data sets are contained in the data archives for each one.

Nutrient Sources

Natural Nutrient Sources

The nutrient source terms representing west-side and east-side forestland in the total nitrogen (TN) and total phosphorus (TP) models and scrubland and grassland in the TP model were used as surrogates for natural sources of nitrogen and phosphorus because direct estimates of these natural nutrient sources were not available for the Pacific Northwest (PNW). These natural nutrient sources result from complex biological and abiotic processes. In the TN model the natural source of nitrogen was fixation of atmospheric nitrogen in forests and in the TP model the natural source of phosphorus was the weathering of geologic material. These nutrient source terms were derived from the 2001 National Land Cover Data Base (NLCD) (Homer and others, 2004) (Michael Wieczorek, U.S. Geological Survey, written commun., June 11, 2011). The total areas of forestland (NLCD categories 41, 42, and 43), scrubland (NLCD category 52), and grassland (NLCD category 71) were summarized for each incremental NHD catchment.

Point Sources

The nutrient source terms representing point sources included 785 facilities having a National Pollutant Discharge Elimination System (NPDES) permit. The discharge of TN and TP for 2002 from sewage treatment and food processing facilities were estimated using measured flow at plant outflows and either onsite measurements of TN and TP or a regional average for a specific industrial classification. The discharge of TN and TP were not estimated for facilities without adequate flow data. The discharges of TN and TP from fish farms and hatcheries for 2002 were estimated by using a mass balance of annual fish production and feed usage and an estimate of nitrogen and phosphorus content for fish and feed (Hal Michael, Washington Department of Fish and Wildlife, oral commun., December 2009). The discharge of TN and TP were not estimated for hatcheries and other aquaculture facilities without adequate production and feed data. The total amount of TN and TP discharged during 2002 from all point sources was estimated for each incremental NHD catchment (see

GIS data #14 and #22). The data set of point sources used in the NHD SPARROW model included more facilities than the one used in the RF1 SPARROW model (785 versus 391) (Maupin and Ivahnenko, 2011) and the additional facilities were mostly non-major sewage treatment plants, fish farms and hatcheries, and food processing operations.

Developed Land

The nutrient source term representing developed land was used as surrogate for various nutrient sources originating from residential, commercial, and industrial land. The use of this source was intended to capture non-point sources of nutrients, including fertilizer, animal waste, and failing sewer systems from these settings. However, the atmospheric deposition of nitrogen or nutrients discharged from permitted point sources were not intended to be represented by this source. Developed land was equal to the summed areas of NLCD categories 21, 22, 23, and 24, minus areas representing roads. Areas representing roads were removed for three reasons: (1) forested catchments with extensive logging road networks would incorrectly exhibit a strong developed land signal, (2) an accurate representation of residential land use was needed to estimate the distribution of non-sewered population and non-farm fertilizer (see section Farm Fertilizer and Non-Farm Fertilizer), and (3) the lack of nutrient use along most roadways. The total area of developed land was summarized for each incremental NHD catchment (GIS data #6 and #19). This approach was different from the approach used for the RF1 SPARROW models, which included NLCD categories 22, 23, and 24, but did not explicitly eliminate roads.

Non-Sewered Population

The nutrient source term representing non-sewered population (the number of people not served by a municipal wastewater-treatment plant) was used as a surrogate for nitrogen leaching from septic tanks. The extent of non-sewered population was computed by (1) overlaying census blocks polygons from the 2000 United States census on a grid of developed land (see section Developed Land); (2) distributing the population of each census block equally to each developed land grid cell within that census block; and (3) removing any of the grid cells that were within an area served by municipal sewers during 2002 (GIS data #15). The total non-sewered population was summarized for each incremental NHD catchment (GIS data #11 and #24). This approach was different from the one used in the RF1 SPARROW models, which used 1990 census tract data to estimate the population served by septic tanks (U.S. Geological Survey, 2011a).

Atmospheric Nitrogen Deposition

The nutrient source term representing atmospheric nitrogen deposition was based on the results from the U.S. Environmental Protection Agency Community Multiscale Air Quality (CMAQ) model (R.L. Dennis, U.S. Environmental Protection Agency, written commun., June 2011). The CMAQ model provided spatially refined atmospheric deposition data, which reflected orographic barriers and nitrogen-islands, such as cities and farmland, and included estimates of wet and dry oxidized nitrogen deposition as well as wet and dry reduced nitrogen deposition. The CMAQ atmospheric nitrogen data were summarized for each incremental NHD catchment (GIS data #1 and #16). This data set was different from the atmospheric nitrogen deposition data set used in the RF1 SPARROW TN model (Wieczorek and Lamotte, 2013), which was obtained from the National Atmospheric Deposition Program (NADP), because it represented total nitrogen deposition for 2002 (wet, dry, oxidized, and reduced), whereas the NADP data only represented wet deposition.

Farm Fertilizer and Non-Farm Fertilizer

The nutrient source terms representing farm and non-farm fertilizer were based on county-level estimates of nitrogen and phosphorus from fertilizer use on farmland and non-farm land. The county-level estimates of 2002 farm and non-farm fertilizer use were compiled by the USGS based on statewide fertilizer sales and county-level expenditures (Gronberg and Spahr, 2012). The county-level USGS data for Idaho, Oregon, and Washington, however, were scaled to reflect differences between the statewide sales data compiled by the USGS and data compiled by those states. The statewide sales data were obtained from the Idaho State Department of Agriculture (Richard Killebrew, written commun., March 2011), the online archives of the Oregon State Library (Oregon Department of Agriculture, 2011), and the Washington State Department of Agriculture (Lizette Beckman, written commun., March 2011). The county-level estimates of nitrogen and phosphorus from farm fertilizer were disaggregated equally to NLCD farmland in each county (NLCD categories 81 and 82) and the county-level estimates of nitrogen and phosphorus from non-farm fertilizer use were disaggregated equally to the developed land (see section Developed Land) in each county. The total amount of nitrogen and phosphorus from farm and non-farm fertilizer applied during 2002 was summarized for each incremental NHD catchment (GIS data #2–5 and #18). This approach was different from the approach used for the RF1 SPARROW models (Wieczorek and Lamotte, 2013), which did not account for differences between USGS and state estimates of sales data.

Livestock Manure

The nutrient source terms representing confined cattle and grazing livestock manure were based on the animal populations at registered dairies and feedlots, county-level livestock populations, and animal-specific nutrient generation factors (nitrogen and phosphorus generated per animal per year).

Confined Cattle Manure

The nutrient source term representing confined cattle manure was estimated by multiplying the number of cattle at each dairy or feedlot by the nitrogen and phosphorus generation factors for dairy cows and feedlot cattle, respectively. The location and population of each dairy and feedlot for 2002 were determined from permitting and inspection records obtained from the Idaho, Oregon, and Washington Departments of Agriculture (J.P. Bilderback, written commun., December 2009; Melissa Boschee, written commun., December 2009; and Virginia Prest, written commun., December 2009; respectively). The nitrogen and phosphorus generation factors were obtained from the USGS (D.K. Mueller, written commun., June 2010). All manure generated by a dairy or feedlot was assumed to be applied within the incremental NHD catchment containing that dairy or feedlot. This assumption was based on the facts that (1) most confined operations grow some or most of their own feed on nearby farmland and (2) manure is a low-cost alternative to commercial fertilizer for those crops if it is not trucked long distances (Araji and others, 2001, Sanford and others, 2009). The authors recognize that the assumption of applying all manure locally may over-represent manure applications in some areas because manure at some facilities is not applied locally, may be composted, or is used for energy production. The data needed to quantify these practices, however, does not exist on a regional basis and these practices are likely applied to a small fraction of the total manure generated in the PNW. The total amount of nitrogen and phosphorus generated at dairies and feedlots during 2002 was summarized for each incremental NHD catchment (GIS data #13 and #21). The approach used to develop the nutrient source term representing manure from confined cattle for the NHD SPARROW models differed from the approach used for the RF1 SPARROW models (Wieczorek and Lamotte, 2013). The data set for the RF1 model was based on livestock confinement factors for each county that were obtained from survey data from the 1980s (Kellog and others, 2000) rather than 2002 estimates of cattle populations at individual dairies and feedlots.

Grazing Livestock Manure

The nutrient source term representing grazing livestock manure included manure generated by rangeland cattle and the manure generated by all other non-cattle, non-poultry livestock. Each of these nutrient sources were compiled at the county level and disaggregated to the land based on assumed land use patterns. The manure generated by poultry operations was not included in this nutrient source term because poultry manure typically was not applied to farmland in the county where it was produced. Rather, poultry manure was most often sold statewide and regionally as fertilizer for farm and for nonfarm use (Julie Walker, Washington State Department of Agriculture, oral commun., January 2010), and could not be traced to a particular place of application.

The county-level estimates of nitrogen and phosphorus from rangeland cattle manure was estimated by subtracting the amount of nitrogen and phosphorus generated by dairy and feedlot cattle in each county from the total amount generated by all cattle. The county-level estimates of total of nitrogen and phosphorus generated by all cattle in each county were obtained from the USGS (D.K. Mueller, written commun., June 2010) and were based on 2002 animal counts compiled by the U.S. Department of Agriculture and annual nitrogen and phosphorus generation factors. The county-level estimates of nitrogen and phosphorus from non-cattle grazing livestock manure, which was generally a small percentage of the total grazing livestock manure, were also obtained from the USGS and based on 2002 animal counts compiled by the U.S. Department of Agriculture and annual nitrogen and phosphorus generation factors.

The county-level estimates of nitrogen and phosphorus from rangeland cattle manure were disaggregated equally to the potential grazing land in each county. A raster data set of potential rangeland was compiled using NLCD categories 41 (deciduous forest), 42 (evergreen forest), 43 (mixed forest), 52 (shrub/scrub), 71 (grassland/herbaceous), and 81 (pasture/hay). Potential rangeland was removed if it did not meet criteria for maximum slope, maximum canopy cover, minimum distance to perennial surface water or wells (David Ganskopp, U.S. Department of Agriculture, Agricultural Research Service, oral commun., January 2010), or if the land was managed by the Bureau of Land Management but was not part of a grazing allotment in 2002. The total amount of nitrogen and phosphorus from rangeland cattle manure during 2002 was summarized for each incremental NHD catchment (GIS data #9, #10, and #20). The county-level estimates (2002) of nitrogen and phosphorus from non-cattle grazing livestock manure were disaggregated equally to grassland and pasture in each county (NLCD categories 70 and 81, respectively). The total amount of nitrogen and phosphorus from non-cattle grazing livestock manure during 2002 was summarized for each incremental NHD catchment (GIS data #7, #8, and #20).

The approach used to develop the nutrient source term representing manure from grazing livestock for the NHD SPARROW models differed from the method used for the RF1 SPARROW models (Wieczorek and Lamotte, 2013) primarily in the way that the manure was disaggregated to the land. In the RF1 models, the land that received grazing livestock manure included cultivated crops even though very little of that land is used for livestock grazing in the PNW (U.S. Department of Agriculture, 2009). Additionally, the landscape types that received grazing livestock manure in the RF1 models omitted much of the rangeland throughout the PNW by not including shrub/scrubland and forestland.

Nitrogen Leaching from Red Alder Trees

The nutrient source term representing the leaching of nitrogen from read alder trees (*Alnus rubra*) was based on a data set showing the distribution of this plant species throughout western Oregon and Washington. The same data set was used for the RF1 SPARROW models (U.S. Geological Survey, 2011b). The total basal area of red alder trees circa 2002 was summarized for each incremental NHD catchment (GIS data #23).

Springs and Power Returns

A substantial proportion of the flow in some streams in the PNW is due to groundwater discharged from large spring complexes and a substantial proportion of the flow in some streams is due to the return of water that is diverted upstream for power generation. Because springs and power returns represent a routing of water that cannot be modeled within the SPARROW framework they were modeled as nutrient point sources. The nutrients in springs and power returns originate from some or all of the other natural and anthropogenic nutrient sources included in the models, but reach waterways in a concentrated manner similar to point sources. There were sufficient discharge and water-quality data to estimate mean annual TN and TP loads for three of the largest spring complexes—Opal Springs on the Crooked River in Oregon, Thousand Springs on the Snake River near Hagerman, Idaho, and Griffith and Waikiki Springs on the Little Spokane River in Washington. The nutrient loads from power returns were estimated in one of two ways. One approach was to use the fraction of streamflow received from an upstream reach (see section Surface-Water Drainage Network) at the point of diversion to estimate the percentage of load that was delivered to the receiving reach (an assumption was made that no nutrients were lost between the two reaches) and apply this percentage to the mean annual load estimated for a calibration station on or near the diversion reach. The other approach was used when no calibration station was on or near the diversion reach. In this case, measurements of discharge and nutrient concentrations for the diversion at the point where it entered a receiving reach were used to estimate mean annual TN and TP loads. The total amount of TN and TP discharged during 2002 from the three spring complexes and the power returns was estimated for each incremental NHD catchment. This nutrient source was not included in the RF1 SPARROW models.

Land-to-Water Delivery

Solar Radiation

A data set of mean annual solar radiation was obtained for the PNW that represented the period between 1991 and 2005 (National Renewable Energy Laboratory, 2011). The average mean annual solar radiation for each incremental NHD catchment was estimated (GIS data #12 and #25).

Arid Land Irrigated Agriculture

Arid land irrigated agriculture was defined as all agricultural land east of the Cascade Range to which irrigation water was applied. The extent of arid land irrigated agriculture was determined by (1) identifying areas of pasture/hay and cultivated crops (NLCD categories 81 and 82, respectively) that were east of the crest of the Cascade Range and (2) removing areas representing non-irrigated wheat from the areas identified in (1). Land containing non-irrigated wheat was identified based on professional understanding of major agricultural cropping patterns in the PNW, research done for this study, and evaluating the NLCD and aerial photography for obvious signs of irrigation such as center pivot circles or irrigation ditches. The percentage of each NHD incremental catchment containing arid land irrigation was estimated (GIS data #17). This data set was a refinement of the one used in the RF1 SPARROW models (Wise and Johnson, 2011).

Metadata Links to GIS Data

GIS Rasters

1. *Total atmospheric deposition of oxidized and reduced nitrogen in the United States Pacific Northwest for 2002*

Wise, D.R.
This spatial data set was created by the U.S. Geological Survey (USGS) to represent the amount of wet and dry deposition of oxidized and reduced nitrogen in the Pacific Northwest region of the United States (Hydro Region 17; MRB7) during 2002.
http://water.usgs.gov/lookup/getspatial?sir2013-5103_PNW_SPARROW_AtmosphericDeposition_td_tn

2. *Application of fertilizer nitrogen to farm land in the United States Pacific Northwest for 2002*

Wise, D.R.
This spatial data set was created by the U.S. Geological Survey (USGS) to represent the amount of fertilizer nitrogen that was applied to farm land in the Pacific Northwest region of the United States (Hydro Region 17; MRB7) during 2002.
http://water.usgs.gov/lookup/getspatial?sir2013-5103_PNW_SPARROW_FertilizerDistribution_Farm_fert_n

3. *Application of fertilizer phosphorus to farm land in the United States Pacific Northwest for 2002*

Wise, D.R.
This spatial data set was created by the U.S. Geological Survey (USGS) to represent the amount of fertilizer phosphorus that was applied to farm land in the Pacific Northwest region of the United States (Hydro Region 17; MRB7) during 2002.
http://water.usgs.gov/lookup/getspatial?sir2013-5103_PNW_SPARROW_FertilizerDistribution_Farm_fert_p

4. *Application of fertilizer nitrogen to nonfarm land in the United States Pacific Northwest for 2002*

Wise, D.R.
This spatial data set was created by the U.S. Geological Survey (USGS) to represent the amount of fertilizer nitrogen that was applied to nonfarm land in the Pacific Northwest region of the United States (Hydro Region 17; MRB7) during 2002.
http://water.usgs.gov/lookup/getspatial?sir2013-5103_PNW_SPARROW_FertilizerDistribution_Nonfarm_nfert_n

5. *Application of fertilizer phosphorus to nonfarm land in the United States Pacific Northwest for 2002*

Wise, D.R.
This spatial data set was created by the U.S. Geological Survey (USGS) to represent the amount of fertilizer phosphorus that was applied to nonfarm land in the Pacific Northwest region of the United States (Hydro Region 17; MRB7) during 2002.
http://water.usgs.gov/lookup/getspatial?sir2013-5103_PNW_SPARROW_FertilizerDistribution_Nonfarm_nfert_p

6. *Nonroad developed land in the United States Pacific Northwest for 2001*

Wise, D.R.
This spatial data set was created by the U.S. Geological Survey (USGS) to represent the extent of non-road developed land in the Pacific Northwest region of the United States (Hydro Region 17; MRB7) in 2002.
http://water.usgs.gov/lookup/getspatial?sir2013-5103_PNW_SPARROW_LandUseDistribution_dev_nonroad

7. *Application of nitrogen generated by non-cattle livestock to pasture land in the United States Pacific Northwest for 2002*

Wise, D.R.

This spatial data set was created by the U.S. Geological Survey (USGS) to represent the amount of nitrogen generated by pastured, non-cattle livestock that was applied to pasture land in the Pacific Northwest region of the United States (Hydro Region 17; MRB7) during 2002. http://water.usgs.gov/lookup/getspatial?sir2013-5103_PNW_SPARROW_LivestockWasteDistribution_Pasture_past_n

8. *Application of phosphorus generated by non-cattle livestock to pasture land in the United States Pacific Northwest for 2002*

Wise, D.R.

This spatial data set was created by the U.S. Geological Survey (USGS) to represent the amount of phosphorus generated by pastured, non-cattle livestock that was applied to pasture land in the Pacific Northwest region of the United States (Hydro Region 17; MRB7) during 2002. http://water.usgs.gov/lookup/getspatial?sir2013-5103_PNW_SPARROW_LivestockWasteDistribution_Pasture_past_p

9. *Application of nitrogen generated by grazing cattle to range land in the United States Pacific Northwest for 2002*

Pischel, E.M.

This spatial data set was created by the U.S. Geological Survey (USGS) to represent the amount of nitrogen generated by grazing cattle that was applied to range land in the Pacific Northwest region of the United States (Hydro Region 17; MRB7) during 2002. http://water.usgs.gov/lookup/getspatial?sir2013-5103_PNW_SPARROW_LivestockWasteDistribution_Rangeland_range_n

10. *Application of phosphorus generated by grazing cattle to range land in the United States Pacific Northwest for 2002*

Pischel, E.M.

This spatial data set was created by the U.S. Geological Survey (USGS) to represent the amount of phosphorus generated by grazing cattle that was applied to range land in the Pacific Northwest region of the United States (Hydro Region 17; MRB7) during 2002. http://water.usgs.gov/lookup/getspatial?sir2013-5103_PNW_SPARROW_LivestockWasteDistribution_Rangeland_range_p

11. *Nonsewered population in the United States Pacific Northwest for 2002*

Wise, D.R.

This spatial data set was created by the U.S. Geological Survey (USGS) to represent populations that did not have access to municipal sewer service in the Pacific Northwest region of the United States (Hydro Region 17; MRB7) in 2002. http://water.usgs.gov/lookup/getspatial?sir2013-5103_PNW_SPARROW_SewageTreatment_NonSewered_nonsewered

12. *Mean annual solar radiation in the United States Pacific Northwest (1991–2005)*

Wise, D.R.

This spatial data set was created by the U.S. Geological Survey (USGS) to represent mean annual solar radiation in the Pacific Northwest region of the United States (Hydro Region 17; MRB7) for the period between 1991 and 2005. http://water.usgs.gov/lookup/getspatial?sir2013-5103_PNW_SPARROW_SolarRadiation_ghi_100

GIS Shapefiles

13. *Nutrient generation at dairies and feedlots in the United States Pacific Northwest for 2002*

Wise, D.R.

This spatial data set was created by the U.S. Geological Survey (USGS) to represent the amount of nitrogen and phosphorus generated by cattle at dairies and feedlots in the Pacific Northwest region of the United States (Hydro Region 17; MRB7) during 2002. http://water.usgs.gov/lookup/getspatial?sir2013-5103_PNW_SPARROW_PointSources_ConfinedCattle_CAFOs

14. *Point source nutrient discharges to surface water in the United States Pacific Northwest for 2002*

Wise, D.R.

This spatial data set was created by the U.S. Geological Survey (USGS) to represent the amount of total nitrogen and total phosphorus discharged to surface waters in the Pacific Northwest region of the United States (Hydro Region 17; MRB7) during 2002 from individual permitted wastewater facilities. http://water.usgs.gov/lookup/getspatial?sir2013-5103_PNW_SPARROW_PointSources_NHD

15. *Areas with access to municipal sewer service in the United States Pacific Northwest for 2002*

Wise, D.R.
This spatial data set was created by the U.S. Geological Survey (USGS) to represent areas that had access to municipal sewer service in the Pacific Northwest region of the United States (Hydro Region 17; MRB7) in 2002.
http://water.usgs.gov/lookup/getspatial?sir2013-5103_PNW_SPARROW_SewageTreatment_Sewered_SeweredAreas

Summary Tables

16. *Atmospheric deposition of nitrogen in the United States Pacific Northwest for 2002 summarized for NHDPlus v2 catchments*

Wise, D.R.
This spatial data set was created by the U.S. Geological Survey (USGS) to represent the amount of atmospheric nitrogen deposition in the Pacific Northwest region of the United States (Hydro Region 17; MRB7) during 2002 within each incremental watershed delineated in the NHDPlus v2 dataset.
http://water.usgs.gov/lookup/getspatial?sir2013-5103_PNW_SPARROW_AtmosphericDeposition_summary

17. *Arid land irrigation in the United States Pacific Northwest for 2001 summarized for NHDPlus v2 catchments*

Wise, D.R.
This spatial data set was created by the U.S. Geological Survey (USGS) to represent the area of arid land irrigation in the Pacific Northwest region of the United States (Hydro Region 17; MRB7) during 2001 within each incremental watershed delineated in the NHDPlus v2 dataset.
http://water.usgs.gov/lookup/getspatial?sir2013-5103_PNW_SPARROW_EastsideIrrigation_summary

18. *Fertilizer nutrients applied to farm and nonfarm land in the United States Pacific Northwest for 2002 summarized for NHDPlus v2 catchments*

Wise, D.R.
This spatial data set was created by the U.S. Geological Survey (USGS) to represent the amount of fertilizer nitrogen and phosphorus that was applied to farm and nonfarm land in the Pacific Northwest region of the United States (Hydro Region 17; MRB7) during 2002 within each incremental watershed delineated in the NHDPlus v2 dataset.
http://water.usgs.gov/lookup/getspatial?sir2013-5103_PNW_SPARROW_FertilizerDistribution_summary

19. *Nonroad developed land in the United States Pacific Northwest for 2001 summarized for NHDPlus v2 catchments*

Wise, D.R.
This spatial data set was created by the U.S. Geological Survey (USGS) to represent the area of non-road developed land within each incremental watershed delineated in the NHDPlus v2 dataset in the Pacific Northwest region of the United States (Hydro Region 17; MRB7) in 2001.
http://water.usgs.gov/lookup/getspatial?sir2013-5103_PNW_SPARROW_LandUseDistribution_DevelopedNonRoad_summary

20. *Nutrients generated by livestock applied to farm land, pasture land, and range land in the United States Pacific Northwest for 2002 summarized for NHDPlus v2 catchments*

Wise, D.R.
This spatial data set was created by the U.S. Geological Survey (USGS) to represent the amount of nitrogen and phosphorus that was generated by livestock and applied to land in the Pacific Northwest region of the United States (Hydro Region 17; MRB7) during 2002 within each incremental watershed delineated in the NHDPlus v2 dataset.
http://water.usgs.gov/lookup/getspatial?sir2013-5103_PNW_SPARROW_LivestockWasteDistribution_summary

21. *Nutrient generation at dairies and feedlots in the United States Pacific Northwest for 2002 summarized for NHDPlus v2 catchments*

Wise, D.R.
This spatial data set was created by the U.S. Geological Survey (USGS) to represent the amount of nitrogen and phosphorus generated by cattle at dairies and feedlots in the Pacific Northwest region of the United States (Hydro Region 17; MRB7) during 2002 within each incremental watershed delineated in the NHDPlus v2 dataset.
http://water.usgs.gov/lookup/getspatial?sir2013-5103_PNW_SPARROW_PointSources_ConfinedCattle_CAFO_summary

22. *Point source nutrient discharges to surface water in the United States Pacific Northwest for 2002 summarized for NHDPlus v2 catchments*

Wise, D.R.
This spatial data set was created by the U.S. Geological Survey (USGS) to represent the amount of total nitrogen and total phosphorus discharged to surface waters in the Pacific Northwest region of the United States (Hydro Region 17; MRB7) during 2002 from all permitted wastewater facilities located within each incremental watershed delineated in the NHDPlus v2 dataset.
http://water.usgs.gov/lookup/getspatial?sir2013-5103_PNW_SPARROW_PointSources_NHD_summary

23. *Red alder trees distribution in the United States Pacific Northwest for 2002 summarized for NHDPlus v2 catchments*

Wise, D.R.
This spatial data set was created by the U.S. Geological Survey (USGS) to represent the basal area of red alder trees within each incremental watershed delineated in the NHDPlus v2 dataset in the Pacific Northwest region of the United States (Hydro Region 17; MRB7) in 2001.
http://water.usgs.gov/lookup/getspatial?sir2013-5103_PNW_SPARROW_RedAlder_summary

24. *Nonsewered population in the United States Pacific Northwest for 2002 summarized for NHDPlus v2 catchments*

Wise, D.R.
This spatial data set was created by the U.S. Geological Survey (USGS) to represent the population within each incremental watershed delineated in the NHDPlus v2 dataset in the Pacific Northwest region of the United States (Hydro Region 17; MRB7) that did not have access to municipal sewer service in 2002.
http://water.usgs.gov/lookup/getspatial?sir2013-5103_PNW_SPARROW_SewageTreatment_NonSewered_summary

25. *Mean annual solar radiation in the United States Pacific Northwest (1991–2005) summarized for NHDPlus v2 catchments*

Wise, D.R.
This spatial data set was created by the U.S. Geological Survey (USGS) to represent mean annual solar radiation in the Pacific Northwest region of the United States (Hydro Region 17; MRB7) for the period between 1991 and 2005 within each incremental watershed delineated in the NNHDPlus v2 dataset.
http://water.usgs.gov/lookup/getspatial?sir2013-5103_PNW_SPARROW_SolarRadiation_summary

References Cited

Araji, A.A., Abdo, Z.O., and Joyce, P., 2001, Efficient use of animal manure on cropland–Economic analysis: Bioresource Technology, v. 79, no. 2, p. 179–191.

Gronberg, J.M., and Spahr, N.E., 2012, County-level estimates of nitrogen and phosphorus from commercial fertilizer for the conterminous United States, 1987–2006: U.S. Geological Survey Scientific Investigations Report 2012-5207, 30 p.

Homer, Collin, Huang, Chengquan, Yang, Limin, Wylie, Bruce, and Coan, Michael, 2004, Development of a 2001 national land-cover database for the United States: Photogrammetric Engineering and Remote Sensing, v. 70, no. 7, p. 829–840.

Kellogg, R.L., Lander, C.H., Moffitt, D.H., and Gollehon, Noel, 2000, Manure nutrients relative to the capacity of cropland and pastureland to assimilate nutrients—Spatial and temporal trends for the United States: U.S. Department of Agriculture, Natural Resources Conservation Service, accessed March 17, 2013, at http://www.nrcs.usda.gov/wps/portal/nrcs/detail/national/technical/nra/rca/?&cid=nrcs143_014126

Maupin, M.A., and Ivahnenko, Tamara, 2011, Nutrient loadings to streams of the continental United States from municipal and industrial effluent: Journal of the American Water Resources Association, v, 47, no. 5, p. 950–964.

National Renewable Energy Laboratory, 2011, PV Solar Radiation (10 km)—Static Maps: National Renewable Energy Laboratory Web site, accessed March 5, 2013, at http://www.nrel.gov/gis/solar.html.

Oregon Department of Agriculture, 2011, 2005 Oregon tonnage summary: Salem, Oreg., Department of Agriculture Fertilizer Program, accessed March 5, 2013, at http://library.state.or.us/repository/2006/200610061405575/tonnage05_2.pdf.

Sanford, G.R., Posner, J.L., and Hadley, G.L., 2009, Economics of hauling dairy slurry and its value in Wisconsin corn grain systems: Journal of Agricultural, Food, and Environmental Sciences, v. 3, no. 1, p. 1–10.

U.S. Department of Agriculture, 2009, Total grazing land, by region and States, United States, 2007: U.S. Department of Agriculture, National Agricultural Statistics Service, accessed March 18, 2013, at http://www.ers.usda.gov/datafiles/Major_Land_Uses/Summary_tables/Summary_Table_4_total_grazing_land_by_region_and_state_2007.xls.

U.S Geological Survey, 2011a, Location of septic sewer systems in the Pacific Northwest: U.S. Geological Survey, Water Resources NSDI Node Web site, accessed March 5, 2013, at http://water.usgs.gov/GIS/metadata/usgswrd/XML/septicSystems_MRB7.xml.

U.S Geological Survey, 2011b, Red alder basal area, by stream reach, for the Pacific Northwest: U.S. Geological Survey, Water Resources NSDI Node Web site, accessed March 5, 2013, at http://water.usgs.gov/GIS/metadata/usgswrd/XML/redAlderSources_MRB7.xml.

Wieczorek, M.E., and Lamotte, A.E., 2013, Attributes for MRB_E2RF1 Catchments by major river basins in the conterminous United States. U.S. Geological Survey Digital Data Series DS-491, accessed March 18, 2013 at http://water.usgs.gov/nawqa/modeling/rf1attributes.html.

Wise, D.R., and Johnson, H.M., 2011, Surface-water nutrient conditions and sources in the United States Pacific Northwest: Journal of the American Water Resources Association, v. 47, no. 5, p. 1110–1135.

Appendix B.   Model Calibration and Prediction Results

The studentized residuals for the PNW NHD SPARROW models for TN and TP are shown in figures B1 and B2, respectively. The studentized residual is equal to the model residual (the difference between the natural logarithm of measured load and predicted load) divided by an estimate of its standard deviation. The negative values indicate over prediction and the positive values indicate under prediction.

The incremental yields for total nitrogen and total phosphorus in kilograms per hectare per year are shown in figures B3 and B4, respectively. The mean annual total nitrogen and phosphorus loads and yields predicted by the PNW NHD SPARROW models are available online in a tab-delimited ASCII file at http://pubs.usgs.gov/sir/2013/5103/. The file includes predictions for individual stream reaches in the Pacific Northwest as defined by the National Hydrography Dataset Plus (NHDPlus, Pacific Northwest region [17]) medium resolution [1:100,000-scale] geospatial data set (Horizon Systems, 2013). SPARROW prediction variables in the ASCII file are described in the header (denoted by lines starting with "#"), and include COMID (common identifier of an NHD reach), AreaSqKM (area of the incremental NHD catchment, in square kilometers), TotDASqKM (total area draining to a reach, in square kilometers), predictions of the local mean annual load for each reach (in kilograms per year), predictions of the total mean annual load for each reach that is attributable to all upstream nutrient sources (in kilograms per year), and predictions of the total mean annual load for each reach that is attributable to individual upstream nutrient sources (in kilograms per year).

Reference Cited

Horizon Systems, 2013, NHDPlus V2 Data: Horizon Systems database, accessed March 18, 2013, at http://www.horizon-systems.com/nhdplus/.

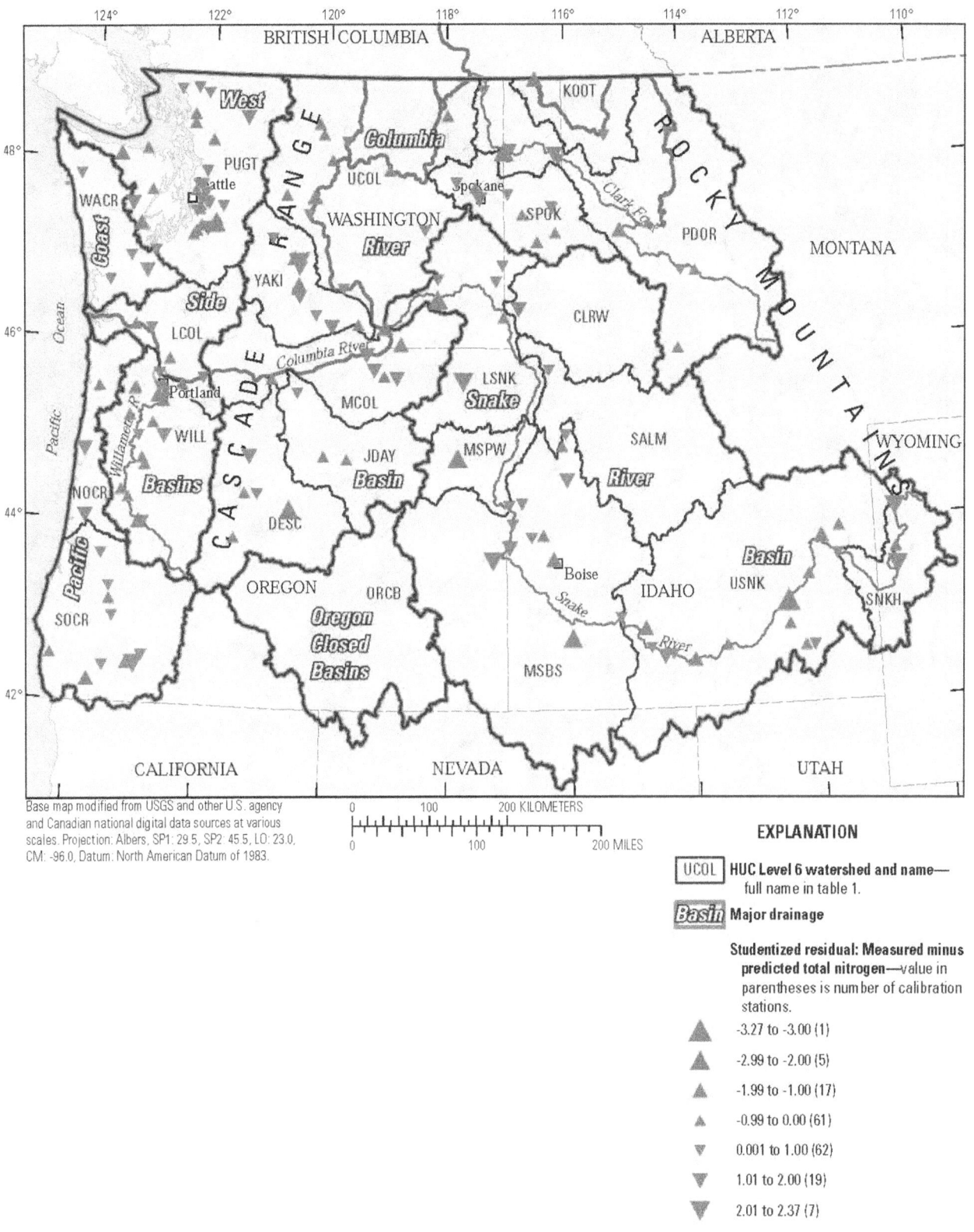

Figure B1.   Spatial distribution of residual stream load for the total nitrogen National Hydrography Dataset SPARROW model developed for the United States Pacific Northwest.

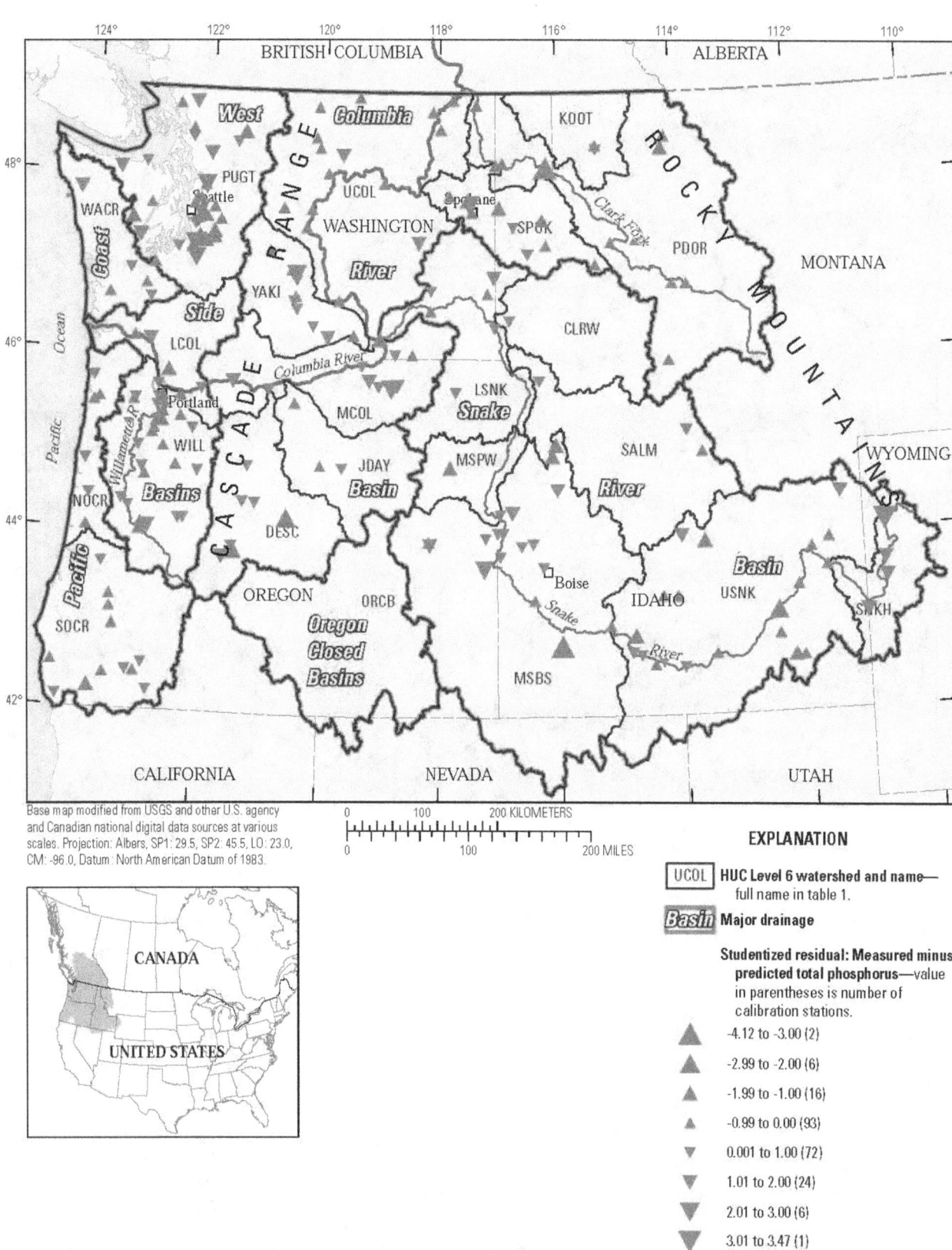

Figure B2. Spatial distribution of residual stream load for the total phosphorus National Hydrography Dataset SPARROW model developed for the United States Pacific Northwest.

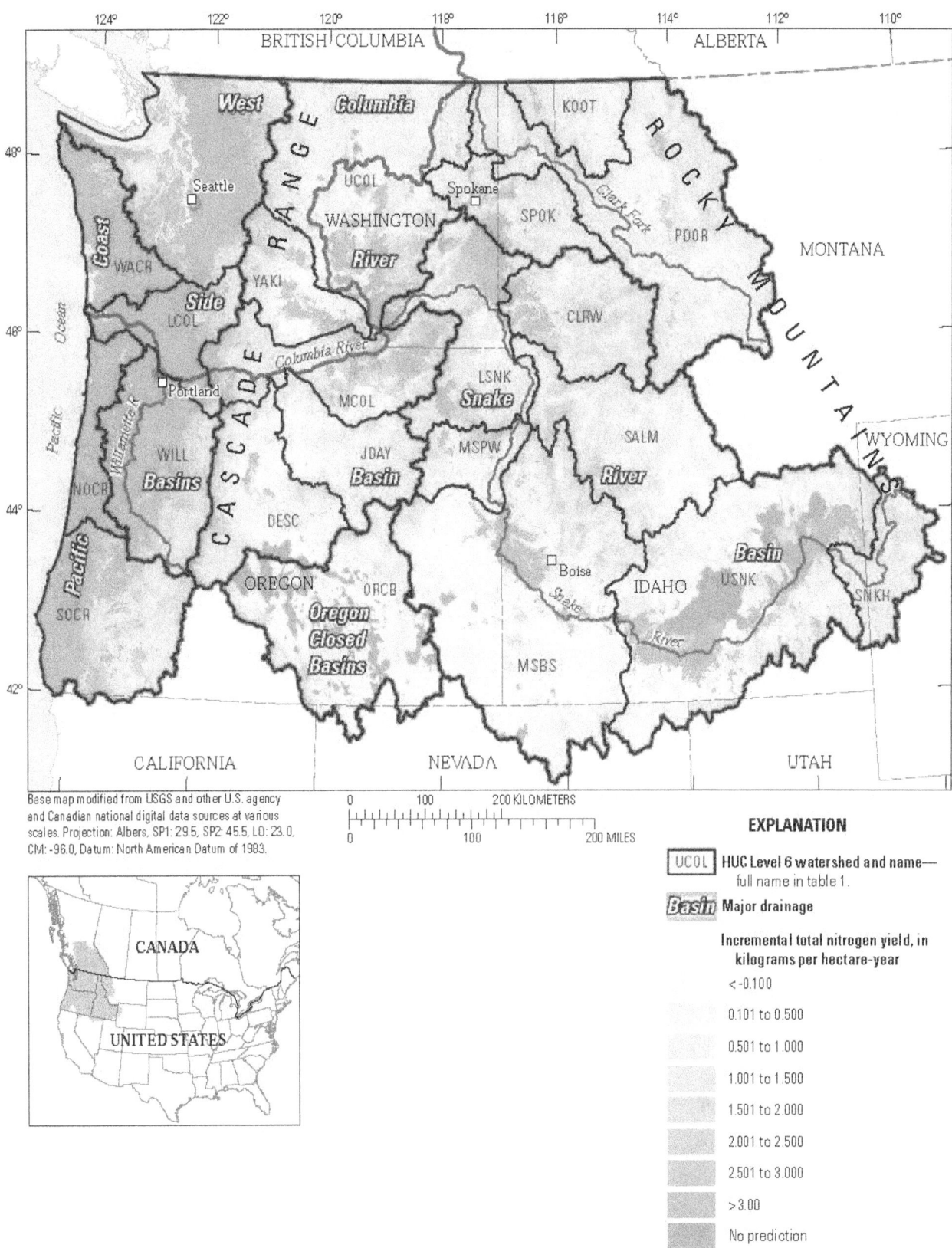

EXPLANATION

UCOL   HUC Level 6 watershed and name—
full name in table 1.

Basin   Major drainage

Incremental total nitrogen yield, in
kilograms per hectare-year

< -0.100

0.101 to 0.500

0.501 to 1.000

1.001 to 1.500

1.501 to 2.000

2.001 to 2.500

2.501 to 3.000

> 3.00

No prediction

Figure B3.   Incremental total nitrogen yields for National Hydrography Dataset catchments in the United States Pacific Northwest (2002 conditions).

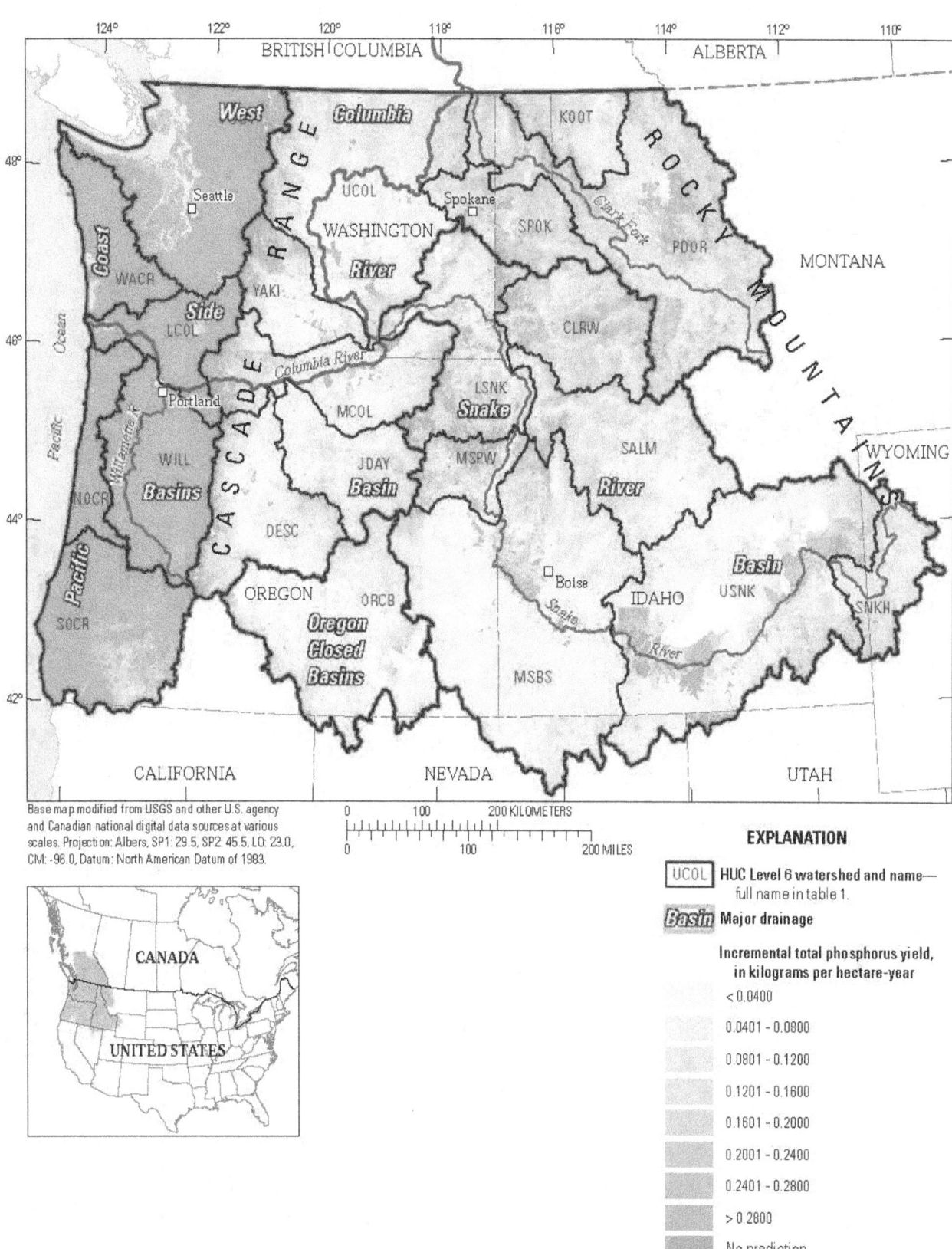

Figure B4. Incremental total phosphorus yields for National Hydrography Dataset catchments in the United States Pacific Northwest (2002 conditions).